On Certain Unitary Representations of an Infinite Group of Transformations

by **Léon Van Hove**

On Certain Unitary Representations of an Infinite Group of Transformations

by **Léon Van Hove**

Translated by
Marcus Berg and Cécile DeWitt-Morette
University of Texas

World Scientific
Singapore • New Jersey • London • Hong Kong

Published by

World Scientific Publishing Co. Pte. Ltd.
P O Box 128, Farrer Road, Singapore 912805
USA office: Suite 1B, 1060 Main Street, River Edge, NJ 07661
UK office: 57 Shelton Street, Covent Garden, London WC2H 9HE

British Library Cataloguing-in-Publication Data
A catalogue record for this book is available from the British Library.

ON CERTAIN UNITARY REPRESENTATIONS OF AN INFINITE GROUP OF TRANSFORMATIONS

Copyright © 2001 by World Scientific Publishing Co. Pte. Ltd.

All rights reserved. This book, or parts thereof, may not be reproduced in any form or by any means, electronic or mechanical, including photocopying, recording or any information storage and retrieval system now known or to be invented, without written permission from the Publisher.

For photocopying of material in this volume, please pay a copying fee through the Copyright Clearance Center, Inc., 222 Rosewood Drive, Danvers, MA 01923, USA. In this case permission to photocopy is not required from the publisher.

ISBN 981-02-4643-9 (pbk)

FOREWORD

In presenting this thesis, the author would like to express his gratitude to the University of Brussels and the professors under which he has had the advantage of studying, for the training he has received and the working conditions he has enjoyed. He would especially like to pay his respects to Professor Lepage who, guiding his early work by example and suggestions, steadily exerted a profound influence on him, and to Professor Géhéniau who opened the paths of research in theoretical physics for him.

The majority of the results presented in the following pages were obtained during the author's stay at the Institute for Advanced Study in Princeton (New Jersey, USA). The author would like to thank the Institute for Advanced Study and its director, Professor J.R. Oppenheimer, for the grant he was offered and for the hospitality that was extended to him. His gratitude also goes to the Government of the United States of America which covered his travel expenses through a Fulbright grant, and to the University of Brussels which granted him the necessary leave.

In preparing this work the author benefited greatly from discussions with a number of people. Among them he would especially like to thank Drs. J. von Neumann, V. Bargmann and I. E. Segal for their continued interest in his work, Drs. O. Klein and W. Pauli who showed interest in certain of his results from a more physical point of view, and Drs. Th. Lepage and J. Géhéniau whose suggestions allowed him to improve the writing of this thesis.

Free University of Brussels   April 20, 1951.

FOREWORD TO THE TRANSLATION

On April 20, 1951, Leon Van Hove presented his thesis "Sur certaines représentations unitaires d'un groupe infini de transformations" to the *Université libre de Bruxelles* (Free University of Brussels). On April 18, 1951, the University of Grenoble approved the creation of *L'Ecole d'été de physique théorique* at Les Houches (Haute Savoie, France). The first session of the "Ecole des Houches" began on July 15, 1951 with a month-long course by Van Hove on quantum mechanics. The lecture notes for this course were written for the benefit of physicists who—like most of their colleagues outside the US, Canada, and England at that time—did not know quantum mechanics, but wanted to learn it seriously. Van Hove's course met their expectations fully. The physics course benefitted from the mathematical expertise of the lecturer, which is also apparent in the present thesis. Without his own research as a scaffolding, Van Hove could not have built the short and beautiful course which provided the participants with a solid, useful foundation in modern physics.

The lecture notes are in French. If they had been in English they would have been published together with the thesis translation. The first three pages of the notes are reproduced at the end of this monograph. The set of notes was reproduced by stencils and distributed to the participants at the beginning of the course.

The translation of the thesis was initiated in late 2000, when Bob Hermann, formely in the Department of Mathematics at MIT, sent to Leon Van Hove's son Michel his view on Leon's thesis: "I would consider it as one of the most important mathematical physics papers of the past fifty years, containing the key ideas for what has become known as 'geometric quantization'.". Indeed, the thesis is interesting both to historians of science and to theoretical physicists and mathematicians exploring the relationships between quantum and classical physics, based on the Hilbert-space approach to classical mechanics.

To appreciate the historical interest of Van Hove's work in the late 1940s and early 1950s, one recalls that in those days physicists and mathematicians, apart from a few exceptions, did not talk to each other. It was not until the late 1960s that they began to interact profitably. Often some mathematical requirements seem superfluous to a phycisist—an attitude which can be useful but may also be dangerous. The reason for this attitude is that exciting mathematical progress is often presented in the literature in a sequence of definitions, lemmas, theorems which appear dry to a non-specialist. Leon Van Hove does not water down the mathematics, but he is not dry. For instance, in his hands, a density theorem becomes exciting.

To help the reader appreciate the current scientific interest of Van Hove's thesis, we have prepared an abstract, found at the end of this monograph.

Traduttore, traditore. Marcus Berg, the translator, and Cécile DeWitt-Morette, the supervisor, have talked over a number of alternatives and tried to choose the ones that Leon Van Hove would have preferred. We have put translator's notes at the end, marked in the text by notes such as this: ^{t2} Any translation can, of course, always be more polished, but we are happy with its current state and we wish to have it published for the fiftieth anniversary of the School at Les Houches.

>
> Marcus Berg
> Cécile DeWitt-Morette
> Bob Hermann

March 9, 2001.

Table of Contents

FOREWORD v

CHAPTER I: Introduction and Review. 1
   1. Introduction 1
   2. Review of Hypermaximally Symmetric Operators 6

CHAPTER II: The Group of Transformations That Leaves Invariant
      the Pfaff Form $ds - \sum p_j dq_j$. 10
   3. Definition of the Group Γ and Immediate Properties 10
   4. Some Subgroups of Γ 12
   5. Infinitesimal Transformations in the Group Γ 14
   6. A Lie Algebra of Infinitesimal Transformations in Γ 18

CHAPTER III: Unitary Representations $\mathcal{R}$ and $\mathcal{R}^{(\alpha)}$ of the Group Γ. 21
   7. The Representation $\mathcal{R}$ 21
   8. The Representations $\mathcal{R}^{(\alpha)}$ 22
   9. Reduction of the Representation $\mathcal{R}$ 23
   10. Linear Manifolds Invariant Under the Representation $\mathcal{R}$ 25

CHAPTER IV: Infinitesimal Transformations in the
      Representations $\mathcal{R}$ and $\mathcal{R}^{(\alpha)}$. 28
   11. Infinitesimal Transformations in the Representations $\mathcal{R}$ 28
   12. Infinitesimal Transformations in $\mathcal{R}^{(\alpha)}$.
       Properties of the Obtained Operators 30
   13. A Theorem on Infinitesimal Transformations of a
       Manifold Into Itself 32
   14. Application to Operators $H[f]$, $H^{(\alpha)}[f]$ 40
   15. A Lie Algebra of Operators 41

CHAPTER V: Irreducibility of the Representations $\mathcal{R}^{(\alpha)}$. 44
   16. New Expression for the Operators $H^{(\alpha)}[f]$ for $\alpha \neq 0$ 44
   17. Reduction of the Representation $\mathcal{R}^{(\alpha)}$ of the
       Subgroup L for $\alpha \neq 0$ 47
   18. Irreducibility of the Representation $\mathcal{R}^{(\alpha)}$ of the
       Group Γ for $\alpha \neq 0$ 51
   19. Irreducibility of the Representation $\mathcal{R}^{(0)}$ 52

CHAPTER VI: Comparison Between Quantum Mechanical
and Classical Operators 55
   20. The Problem of Relations Between Quantum and
       Classical Descriptions of a System 55
   21. Characterization of Operators in Quantum Mechanics 56
   22. Comparison of the Operators $H^{(\alpha)}[f]$ with the
       Operators of Quantum Mechanics 59
   23. Difficulties with a Unique Correspondence Between
       Classical and Quantum Quantities 61
   24. Bijective Correspondence Between Quadratic Quantities .. 71
   25. Correspondence for Common Quantities 79

REFERENCES 85

CHAPTER I

Introduction and Review.

1 Introduction

On the space (s, p, q) of $2n + 1$ real variables $s, p_1, \ldots, p_n, q_1, \ldots, q_n$, $(-\infty < s < +\infty, -\infty < p_j < +\infty, -\infty < q_j < +\infty)$, we define the Pfaff form[t3]

$$\bar{\omega} = ds - \sum_1^n p_j dq_j \qquad (1.1)$$

and we consider the group Γ of bijective, infinitely differentiable transformations of the space (s, p, q) onto itself, which leave the form $\bar{\omega}$ invariant.

The transformations in Γ leave invariant the volume element $ds\, dp_1 \ldots dp_n\, dq_1 \ldots dq_n$. Therefore, one may immediately construct a representation $\mathcal{R}$ of Γ by unitary transformations of the Hilbert space of measurable and square integrable functions $\phi(s, p_1, \ldots, p_n, q_1, \ldots, q_n)$. For every element γ of Γ, there is a unitary operator U_γ satisfying the equation[t4]

$$U_\gamma \phi(\Omega) = \phi(\gamma^{-1}\Omega)$$

where Ω denotes an arbitrary point in the space (s, p, q). Our goal is to study the representation $\mathcal{R}$. We treat the following two problems:

1° *Reducibility*: we reduce $\mathcal{R}$ to its irreducible components $\mathcal{R}^{(\alpha)}$; the $\mathcal{R}^{(\alpha)}$ are also unitary representations of the Hilbert space; they depend on the parameter α which can take on any real value.

2° *Representation of infinitesimal transformations*: We introduce systems of hypermaximally antisymmetric[t5] operators representing, in $\mathcal{R}$ and $\mathcal{R}^{(\alpha)}$, infinitesimal transformations of the group Γ; we show how it is possible to endow these systems of operators with the algebraic structure of the set of infinitesimal transformations in Γ. We note in passing that this algebraic structure is not simple (we are not dealing with a Lie algebra) due to the nature of infinite-dimensional[t6] group (in the Lie sense) of Γ and the noncompact character of the space (s, p, q).

Furthermore, we indicate and discuss a remarkable analogy between the operators of Quantum Mechanics and the operators introduced by the representation $\mathcal{R}^{(\alpha)}$ for $\alpha = 2\pi/h$, where h is Planck's constant.

Since Sophus Lie, groups of transformations that leave the Pfaff form $\bar\omega$ invariant have often been studied. However, these groups differ from our group Γ since they consist of locally defined transformations, in general in the neighborhood of a point; in addition, their transformations are often assumed to be analytic. In contrast, placing ourselves with a global point of view, we require our transformations to map bijectively the entire space (s,p,q) into itself; we assume them to be C^∞, which is the weakest differentiability hypothesis that locally guarantees the existence of commutators for infinitesimal transformations in the group. Moreover, this hypothesis also seems essential for the validity of our results concerning the representation of infinitesimal transformations in $\mathcal{R}$, $\mathcal{R}^{(\alpha)}$.

Since the transformations leaving the form $\bar\omega$ invariant depend on arbitrary functions, Γ is an infinite-dimensional continuous group. As opposed to finite-dimensional continuous groups, for which the topological structure is well-known (they are compact or locally compact, see for example Chevalley [5a]), infinite-dimensional continuous groups of transformation have escaped a satisfactory topological study until now (see Birkhoff [2] p. 62): one has not been able to give them topological structure properly adapted to their algebraic group structure and to the existence of infinitesimal transformations. The representations $\mathcal{R}$, $\mathcal{R}^{(\alpha)}$ are sufficiently simple to be studied without knowledge of the topological structure of the group Γ. It even seems that, given the current development of Hilbert space techniques, these representations could provide a new method for improving our understanding of the structure of Γ; moreover, such a method would be applicable to most infinite-dimensional continuous groups of transformation. The present work does not approach these general problems, however.

Our interest in the group Γ originates mainly in its fundamental importance, and the importance of some of its subgroups, for Classical Mechanics and for the passage from Classical Mechanics to Quantum Mechanics (quantization problem). It is known that in Classical Mechanics, the various possible motions of a dynamical system[b] are obtained in the Calculus of Variations by extremizing a simple integral, namely the action.[t7] To such an integral one

---

[a] Bibliographical references are at the end of this work, ordered by author name and year of publication.
[b] Following the terminology of Whittaker [35], we call dynamical system the physical object for which motion is described, in various approximations, by Classical Mechanics or Quantum Mechanics. For example, we may consider a collection of interacting particles subject to given external forces. Here we only discuss dynamical systems with a finite number of degrees of freedom.

associates a Pfaff form which is ordinarily written

$$\omega = \sum_1^n p_j dq_j - h\, dt \qquad (1.2)$$

where t denotes time, p_j, q_j the canonically conjugate pairs of variables and $h = h(p,q)$ the Hamiltonian (cf. Whittaker [35], pp. 307). The critical points of the action (i.e. its extrema) are given by the equations of the characteristic system of the exterior differential $d\omega$ (see Cartan [4] or Goursat [12]).[t8] The form ω allows one to define a group closely connected to the dynamical system under consideration: it is the group of transformations of the space of variables t, p_j, q_j into itself such that ω is invariant. This group enjoys the remarkable property that its infinitesimal transformations are in bijective correspondence with the first integrals of the equations of motion, Lie brackets corresponding to Poisson brackets. Let us note that it is *not* this group which is normally discussed in treatises of Classical Mechanics, but rather its quotient group by its center C.[t9] This quotient is essentially the group of canonical transformations, and unfortunately it is no longer bijective.[t10]

In regular cases the form ω is of maximum degree $2n + 1$; according to a famous theorem of Pfaff [26], there exists, then, in the neighborhood of every point, a change of variables taking ω into canonical form (1.1). In the simplest case where a unique change of variables gives ω the form (1.1) in the entire space, the group associated to the dynamical system can be defined as being the group Γ. This shows the relation of the group Γ with Classical Mechanics.

The unitary representations $\mathcal{R}$, $\mathcal{R}^{(\alpha)}$ of Γ are also closely related to certain recent developments in Classical Mechanics, inaugurated by Koopman and von Neumann in their research on ergodic theory (see Koopman [15], von Neumann [24] and for ergodic theory as a whole, Hopf [14]). They generalize a method introduced by Koopman for using operators on Hilbert space in Classical Mechanics. Let us briefly show that these operators can be derived from our representations $\mathcal{R}$, $\mathcal{R}^{(\alpha)}$.[t11] Given the expression (1.1) of the form $\bar{\omega}$, the first integrals of the characteristic equations of the form $d\bar{\omega}$ are functions $f(p_1, \ldots p_n, q_1, \ldots q_n)$; restricted to be infinitely differentiable and having suitable behavior at infinity,[c] these functions are in bijective correspondence with infinitesimal transformations in Γ, Lie brackets corresponding to Poisson brackets. In each of the representations $\mathcal{R}$, $\mathcal{R}^{(\alpha)}$, the infinitesimal transformation corresponding to a function $f(p,q)$ is represented by an operator $-iH[f]$, $-iH^{(\alpha)}[f]$ where $H[f]$, $H^{(\alpha)}[f]$ are hypermaximally symmetric;

---

[c] In the following chapter, the family of functions f satisfying these restrictions is precisely defined. This family will be denoted $\mathcal{F}_\Gamma$.

for every f, the operator $H[f]$ is the direct sum of operators $H^{(\alpha)}[f]$. This being the case, the operator used by Koopman and von Neumann in the study of a conservative dynamical system is none other than $H^{(0)}[h]$ where $h(p,q)$ is the Hamiltonian of the system. In fact, Koopman and von Neumann use the representation $\mathcal{R}^{(0)}$ for the one-parameter subgroup which represents the evolution of the dynamical system in time; moreover, they do this by considering canonical transformations, that is transformations in the group Γ/C and not Γ.[d] Our study departs from the works of Koopman and von Neumann, and from those which followed them, in the following respects:

1° Appealing to the the group Γ rather than Γ/C, we obtain instead of the single representation $\mathcal{R}^{(0)}$ the continuous family of representations $\mathcal{R}^{(\alpha)}$, $(-\infty < \alpha < +\infty)$.

2° Instead of studying individual properties of operators $H^{(\alpha)}(f)$, we are interested in the algebraic structure of the family of operators for every value of α, and we show how it reproduces the algebraic structure of the family of infinitesimal transformations in Γ.

It is well-known that Quantum Mechanics uses operators on Hilbert space in an essential manner. The introduction of similar operators in Classical Mechanics naturally poses the question, already raised by von Neumann [24], of a possible analogy between these operators and the quantum operators. An analogy of this kind does not exist for the operators considered by Koopman and von Neumann; indeed in the representation $\mathcal{R}^{(0)}$, the operators $H^{(0)}[p_j]$, $H^{(0)}[q_j]$ corresponding to conjugate variables p_j, q_j, are commuting, whereas the quantum operators P_j, Q_j satisfy

$$P_j Q_j - Q_j P_j = -\hbar i$$

where $\hbar$ is the Planck constant divided by 2π. On the contrary, a remarkable formal analogy with the representation $\mathcal{R}^{(\alpha)}$ for $\alpha = 1/\hbar$ presents itself: for this value of α, the operators $K[f] = \alpha^{-1} H^{(\alpha)}[f]$ in fact satisfy the commutation relations

$$K[f_1]K[f_2] - K[f_2]K[f_1] = \hbar i K[(f_1, f_2)]$$

where (f_1, f_2) is the Poisson bracket. In particular

$$K[p_j]K[q_j] - K[q_j]K[p_j] = -\hbar i \ .$$

---

[d] $\mathcal{R}^{(0)}$ is not a faithful representation of the group Γ; it represents the center C by the identity transformation of the Hilbert space and thus constitutes a representation of the quotient group Γ/C.

The analogy of these relations with quantum commutation relations is striking.

However, there exist essential differences between the operators $K[f]$ and the quantum operators, which are discussed in the last chapter of the present thesis. These differences raise the question whether other representations of the group Γ could provide operators identical to quantum operators. We establish that this is not the case, at least not if one's attention is confined to representations with properties closely related to those of $\mathcal{R}$, $\mathcal{R}^{(\alpha)}$. Nevertheless, the situation is modified if, instead of the group Γ, we consider the subgroup L formed by linear (non-homogeneous) transformations in Γ on the p_j, q_j: we show that quantum operators linear and quadratic in the P_j, Q_j are obtained by a suitable unitary representation of infinitesimal transformations in L.

The plan of the thesis is as follows.[t12] Chapter II is dedicated to the definition of the group Γ and the study of its infinitesimal transformations. The representations $\mathcal{R}$ and $\mathcal{R}^{(\alpha)}$ are constructed in chapter III where the reduction of $\mathcal{R}$ is also carried out. The families of operators $H[f]$, $H^\alpha[f]$ are studied in the following chapter, where it is shown that they have the same algebraic structure as the family of infinitesimal transformations of Γ. Chapter V establishes the irreducibility of the representations $\mathcal{R}^{(\alpha)}$. This result comes from a study of the reducibility of $\mathcal{R}^{(\alpha)}$ considered as a representation of the subgroup L of Γ; whereas in this respect $\mathcal{R}^{(0)}$ is irreducible, $\mathcal{R}^{(\alpha)}$ for $\alpha \neq 0$ admits two irreducible components. Chapter VI concerns the analogy between the operators $H^{(\alpha)}[f]$ and the operators of Quantum Mechanics.

Finally, let us indicate where the problems treated in the present thesis belong in the general development of mathematical research at the Free University of Brussels. These problems were in part inspired by the long tradition of studying infinite-dimensional groups, particularly in the context of Calculus of Variations, Integral Invariants and Classical Mechanics; in this domain, in addition to the work of Dr. De Donder, one must especially mention Drs. Lepage and Debever, who have extensively developed E. Cartan's point of view. On the other hand, the direction of our work has been influenced by the general interest in unitary representations of discrete and continuous finite-dimensional[t13] groups, and their applications to Quantum Mechanics, the importance of which has often been emphasized by Dr. Géhéniau and which has been the subject of unpublished work. In particular, as will be noted later, it has been useful to us to point out the analogy between the problems posed by the representations $\mathcal{R}$, $\mathcal{R}^{(\alpha)}$ of the group Γ and problems related to the theory of unitary representations of locally compact groups in Hilbert space. This very recent theory made its debut in the fundamental treatises of Gelfand and Neumark [9,10] and of Bargmann [1], devoted to the Lorentz group; at the Free University of Brussels, the thesis of Bargmann was carefully studied by Dr. Daniels; following Gelfand and Neumark, Dr. Tits lectured on the construction of unitary representations of the Lorentz group in four-dimensional space. Within the framework of these diverse activities, the objective of the present thesis can be characterized as follows: we apply to the infinite-dimensional group of Classical Mechanics (group Γ) the techniques of unitary representations in Hilbert space, and observe remarkable analogies between the operators so obtained and the operators of Quantum Mechanics.

In the following section we review some notions about operators on Hilbert space that are essential for the understanding of the last four sections.[t14]

2 Review of Hypermaximally Symmetric Operators

We begin with some definitions (see Stone[30]); $\mathcal{H}$ denotes a separable Hilbert space.

a) *Projection operators*:

A operator transforming every vector of $\mathcal{H}$ into its orthogonal projection on a linearly closed manifold[t15] in $\mathcal{H}$ is called projection operator.

b) *Resolution of the identity*:

In the Hilbert space $\mathcal{H}$, a resolution of the identity is a family of projection operators $E(\lambda)$, depending on the real parameter λ for $-\infty < \lambda < +\infty$ and enjoying the following properties:

$$E(\lambda)E(\mu) = E(\lambda) \quad \text{for } \lambda \leq \mu\,;$$
$$E(\lambda)E(\mu) = E(\mu) \quad \text{for } \lambda > \mu\,;$$
$$E(\lambda + \eta)\varphi \to E(\lambda)\varphi \quad \text{for } \varphi \in \mathcal{H},\, \eta > 0,\, \eta \to 0\,;$$
$$E(\lambda)\varphi \to \varphi \quad \text{for } \varphi \in \mathcal{H},\, \lambda \to +\infty\,;$$
$$E(\lambda)\varphi \to 0 \quad \text{for } \varphi \in \mathcal{H},\, \lambda \to -\infty\,.$$

c) *Hypermaximally symmetric operator*:

Let A be a linear operator in $\mathcal{H}$, and $\mathcal{D}$ its domain of definition, which we briefly call its domain ($\mathcal{D} \subset \mathcal{H}$). Let us denote by (φ, ψ) the scalar product of two elements φ and ψ of $\mathcal{H}$. The operator A is said to be symmetric when the following conditions are fulfilled:

1° $\mathcal{D}$ is dense in $\mathcal{H}$;

2° $(A\varphi, \psi) = (\varphi, A\psi)$ for $\varphi \in \mathcal{D}$, $\psi \in \mathcal{D}$.

Let A be an operator on the domain $\mathcal{D}$, dense in $\mathcal{H}$, then the adjoint A^* on the domain $\mathcal{D}^*$ is the operator defined as follows:[t16]

$$(A\varphi_0, \varphi) = (\varphi_0, A^*\varphi), \quad \forall \varphi_0 \in \mathcal{D}, \quad \forall \varphi \in \mathcal{D}^*.$$

An operator A of domain $\mathcal{D}$ is said to be hypermaximally symmetric (self-adjoint, in the terminology of Stone [30], chap. II) when it is symmetric and identical to its adjoint A^*, that is when

$$\mathcal{D} \equiv \mathcal{D}^*, \ A\varphi = A^*\varphi \text{ for } \varphi \in \mathcal{D}.$$

Let us now state the *spectral representation theorem* for hypermaximally symmetric operators. We write $||\varphi|| = (\varphi, \varphi)^{1/2}$ for $\varphi \in \mathcal{H}$. The theorem affirms the existence of a bijective correspondence between the set of hypermaximally symmetric operators and the set of resolutions of the identity of $\mathcal{H}$, with the following properties: if A is a hypermaximally symmetric operator of domain $\mathcal{D}$, and $E(\lambda)$ the corresponding resolution of the identity, then the domain $\mathcal{D}$ is the set of elements φ of $\mathcal{H}$ for which the Stieltjes integral

$$\int_{-\infty}^{+\infty} \lambda^2 \, d||E(\lambda)\varphi||^2 < \infty;$$

for $\varphi \in \mathcal{D}, \psi \in \mathcal{H}$ we have the equation

$$(A\varphi, \psi) = \int_{-\infty}^{+\infty} \lambda \, d(E(\lambda)\varphi, \psi)$$

which completely determines $A\varphi$. Both integrals above are Stieltjes integrals.

Three distinct types of proof have been given for this fundamental theorem. They are due to J. von Neumann [19], F. Riesz [27] and M. H. Stone [30].

Let us indicate some further notions that will be used frequently in the following. Let A, A' be two operators of domain $\mathcal{D}$, $\mathcal{D}'$. One says that A is an extension of A' and A' a restriction of A when $\mathcal{D}$ contains $\mathcal{D}'$ and $A\varphi = A'\varphi$ for all $\varphi \in \mathcal{D}'$. We thus write $A \supset A'$, $A' \subset A$. A symmetric operator A is said to be essentially hypermaximally symmetric (essentially self-adjoint in the terminology of Stone [30], chap. II) when $A^* = A^{**}$. A^* is thus a hypermaximally symmetric extension of A. It is the only such extension—a fundamental fact for understanding chapter IV—by virtue of the following proposition: *if A is essentially hypermaximally symmetric and if B is a symmetric extension of A, then B is essentially hypermaximally symmetric and the hypermaximally* *symmetric operators A^*, B^* are identical.* Since this proposition is less classical than the other theorems stated in the present paragraph, we give its proof: $A \subset B$ entails $A^* \supset B^*$, $A^{**} \subset B^{**}$; A being essentially hypermaximally symmetric, $A^* \equiv A^{**}$, hence $B^* \subset A^* \subset B^{**}$; B being symmetric, $B \subset B^*$, hence $B^* \supset B^{**}$. Consequently $B^* \equiv B^{**} \equiv A^*$.

An operator A is called antisymmetric (hypermaximally antisymmetric, essentially hypermaximally antisymmetric) when the operator iA is symmetric (hypermaximally symmetric, essentially hypermaximally symmetric). If A is essentially hypermaximally antisymmetric, the only hypermaximally antisymmetric extension of A (and of every antisymmetric extension of A) satisfies $A^{**} \equiv -A^*$.

An operator T defined on $\mathcal{H}$ is called bounded when there exists a positive constant b such that $\|T\varphi\| \leq b\|\varphi\|$ for all $\varphi \in \mathcal{H}$. It can be shown that a symmetric operator for which the domain is the entire space $\mathcal{H}$ is both bounded and hypermaximally symmetric.

To end this review we state a few more well-known theorems of Stone establishing the relation between hypermaximally symmetric operators and one-parameter groups of unitary transformations of $\mathcal{H}$.

Theorem A. If A is a hypermaximally symmetric operator, then unitary transformations $U(\tau) = e^{i\tau A}$ for τ real, $-\infty < \tau < +\infty$, form a group:

$$U(\sigma + \tau) = U(\sigma) \cdot U(\tau), \quad (2.1)$$

and are strongly continuous with respect to τ, that is

$$\|U(\sigma)\varphi - U(\tau)\varphi\| \to 0 \text{ for } \sigma \to \tau, \ \varphi \in \mathcal{H}. \quad (2.2)$$

Let us recall that $e^{i\tau A}$ can be defined using the resolution of the identity $E(\lambda)$ corresponding to A

$$\psi' = e^{i\tau A}\psi \quad \text{if} \quad (\psi', \varphi) = \int_{-\infty}^{+\infty} e^{-i\lambda\tau} d(E(\lambda)\psi, \varphi) \text{ for all } \varphi \in \mathcal{H}.$$

Theorem B. If $U(\tau)$ is a family of unitary transformations depending on the real parameter τ, $-\infty < \tau < +\infty$, satisfying the group property (2.1) and the following measurability property:

$$\begin{array}{l}(U(\tau)\varphi, \psi) \text{ is a measurable function of } \tau \\ \text{for all } \varphi \in \mathcal{H},\ \psi \in \mathcal{H},\end{array} \quad (2.3)$$

there exists one and only one hypermaximally symmetric operator A such that $U(\tau) = e^{i\tau A}$ for all values of τ.

We note that the group property (2.1) and the measurability property (2.3) together imply the strong continuity property (2.2).

Theorem C: If A is a hypermaximally symmetric operator of domain $\mathcal{D}$ and if $U(\tau) = e^{i\tau A}$ for all values of the real parameter τ, one has strong convergence

$$\frac{1}{\tau}[U(\tau) - U(0)]\varphi \to \varphi' \quad \text{for} \quad \tau \to 0$$

under the necessary and sufficient property that

$$\varphi \in \mathcal{D}, \; \varphi' = A\varphi .$$

For these theorems one can consult Stone [29,31] as well as von Neumann [23]. They establish a bijective correspondence between hypermaximally symmetric operators and strongly continuous one-parameter groups of unitary transformations of $\mathcal{H}$. One sees that these groups have hypermaximally antisymmetric operators as infinitesimal transformations.

CHAPTER II

The Group of Transformations That Leaves Invariant the Pfaff Form $ds - \sum p_j dq_j$.

We define the group of transformation which leaves invariant the Pfaff form $ds - \sum p_j dq_j$ and indicate some of its subgroups. We study its infinitesimal transformations and construct a family of infinitesimal transformations with Lie algebra structure.

3 Definition of the Group Γ and Immediate Properties

We consider the differential form

$$\bar{\omega} = ds - \sum_{j=1}^{n} p_j dq_j \qquad (3.1)$$

in the space (s, p, q) of $2n+1$ dimensions of real variables s, p_j, q_j

$$-\infty < s < +\infty, \quad -\infty < p_j < +\infty, \quad -\infty < q_j < +\infty$$
$$(j = 1, \ldots n)$$

We recall that a function is called C^∞ when it is infinitely differentiable at all points of its domain of definition. A pointwise transformation will be called C^∞ whenever the new variables are C^∞ functions of the old ones.

We then consider the family Γ of bijective transformations of the space (s, p, q) onto itself which are C^∞ and leave the differential form (3.1) invariant. The product of two transformations in Γ is in Γ; the identity transformation is in Γ.

Let

$$s' = s'(s, p_j, q_j), \; p' = p'_k(s, p_j, q_j), \; q' = q'_k(s, p_j, q_j)$$
$$(j, k = 1, \ldots n)$$

be the equations of a transformation in Γ. From the invariance of the form $\bar{\omega}$

$$ds' - \sum_{j=1}^{n} p'_j dq'_j = ds - \sum_{j=1}^{n} p_j dq_j , \qquad (3.2)$$

we deduce the invariance of the exterior differential $d\bar{\omega}$ and of exterior products $(d\bar{\omega})^n$, $\bar{\omega} \wedge (d\bar{\omega})^n$; hence[t17]

$$\sum dp'_j dq'_j = \sum dp_j dq_j , \qquad (3.3)$$

$$dp'_1 \ldots dp'_n dq'_1 \ldots dq'_n = dp_1 \ldots dp_n\, dq_1 \ldots dq_n\,, \tag{3.4}$$

$$ds'\, dp'_1 \ldots dp'_n dq'_1 \ldots dq'_n = ds\, dp_1 \ldots dp_n\, dq_1 \ldots dq_n\,. \tag{3.5}$$

Equation (3.5) shows that the Jacobian $\partial(s', p'_k, q'_k)/\partial(s, p_j, q_j)$ is one. The transformation therefore admits a C^∞ inverse, which is also in the family Γ. *Thus the family Γ forms a group.*

Equation (3.4) immediately gives

$$\begin{vmatrix} \dfrac{\partial p'_j}{\partial p_k} & \dfrac{\partial p'_j}{\partial q_k} \\ \dfrac{\partial q'_j}{\partial p_k} & \dfrac{\partial q'_j}{\partial q_k} \end{vmatrix} = 1\,. \tag{3.6}$$

Since the left-hand side of (3.3) cannot contain ds, one has

$$\sum_k \left(\frac{\partial q'_k}{\partial q_j} \frac{\partial p'_k}{\partial s} - \frac{\partial p'_k}{\partial q_j} \frac{\partial q'_k}{\partial s} \right) = 0\,, \quad \sum_k \left(\frac{\partial q'_k}{\partial p_j} \frac{\partial p'_k}{\partial s} - \frac{\partial p'_k}{\partial p_j} \frac{\partial q'_k}{\partial s} \right) = 0\,.$$

Because of (3.6) these equations demand

$$\frac{\partial p'_k}{\partial s} = 0\,, \quad \frac{\partial q'_k}{\partial s} = 0\,. \tag{3.7}$$

If one writes (3.2) in the form

$$d(s' - s) = \sum (p'_j dq'_j - p_j dq_j)\,,$$

one sees from (3.7) that the right-hand side does not contain s. Thus

$$s' = s + \pi(p_j, q_j)$$
$$d\pi = \sum_j (p'_j dq'_j - p_j dq_j)\,.$$

Every transformation of the group Γ is thus of the following type

$$\begin{cases} p'_k = p'_k(p_j, q_j)\,, \\ q'_k = q'_k(p_j, q_j)\,, \end{cases} \tag{3.8}$$

$$s' = s + \pi(p_j, q_j) \tag{3.9}$$

with

$$\sum (p'_j dq'_j - p_j dq_j) = d\pi\,.$$

It defines, by the intermediary (3.8), a bijective transformation of the space (p, q) of variables $p_1, \ldots, p_n, q_1, \ldots q_n$ into itself; this transformation is C^∞

and leaves $\sum dp_j dq_j$ invariant. It constitutes a canonical transformation (see for example Carathéodory [3] p. 78). Conversely, given a transformation (3.8) enjoying these properties, it extends in an infinite number of ways to a transformation of Γ, because the function π is defined only up to a constant.

Let us remark that in the definition of the group Γ we could have considered "sufficiently differentiable" transformations instead of limiting ourselves to C^∞ transformations.[t18] All the preceding properties hold for this larger group. However, later on it will be useful to restrict ourselves to C^∞ transformations. We make this restriction from now on purely for convenience.

The group Γ is a continuous group of transformations with elements depending on arbitrary functions. It is therefore a infinite-dimensional group in the sense of Sophus Lie.[e] We have defined it algebraically, that is, as a set of elements for which a composition law is given. It would be desirable to also give it a structure of a topological group by defining a suitable topology. For infinite-dimensional groups of transformation, it is known that this is a difficult problem, which does not appear to have found a satisfactory solution yet (see Birkhoff[2]). We will not concern ourselves with this problem in the following developments.

4 Some Subgroups of Γ

We now give some subgroups of the group Γ.

Center C.

The group C of transformations

$$s' = s + \sigma \,,\ p'_j = p_j \,,\ q'_j = q_j \,,\ (j = 1, \ldots n)$$

where σ is an arbitrary real constant, is an invariant subgroup of Γ. The quotient group Γ/C is isomorphic to the group of bijective transformations of the space (p, q) of variables $p_1, \ldots, p_n, q_1, \ldots, q_n$ onto itself, which are C^∞ and leave the form $\sum dp_j dq_j$ invariant. These are the transformations which are normally studied in treatises of Mechanics and canonical transformations (see for example Carathéodory [3] p. 78; Carathéodory does not restrict himself to C^∞ transformations defined on the entire space (p, q)). *The subgroup C is the center of the group Γ.* It is actually obvious that the transformations in C commute with all transformations in Γ belonging to C if it commutes with

---

[e] However, it must be noted that Lie considered analytic transformations and not C^∞ transformations.

the transformations in the group T defined below.

Subgroup T.

It is the group of transformations

$$s' = s + \sum \alpha_j q_j + \sigma, \quad p'_j = p_j + \alpha_j, \quad q'_j = q_j + \beta_j, \ (j = 1, \ldots n)$$

where the α_j, the β_j and σ are arbitrary real constants. T contains C. Whereas T is not Abelian, T/C does enjoy this property. T/C is the group of translations of the space (p, q).

Subgroup L.

It is the group of transformations

$$s' = s + \sum_{i,j,k} \left(\frac{1}{2} a_{ij} c_{ik} p_j p_k + \frac{1}{2} b_{ij} d_{ik} q_j q_k + b_{ij} c_{ik} q_j p_k \right) + \sum_{j,k} d_{jk} \alpha_j q_k + \sigma,$$

$$p'_j = \sum_k (a_{jk} p_k + b_{jk} q_k) + \alpha_j,$$

$$q'_j = \sum_k (c_{jk} p_k + d_{jk} q_k) + \beta_j,$$

where the α_j, the β_j, and σ are arbitrary real constants and where the real matrix of rank $2n$ [19]

$$\begin{pmatrix} a_{jk} & b_{jk} \\ c_{jk} & d_{jk} \end{pmatrix}$$

is symplectic. Thus it satisfies the following conditions

$$\sum_i (a_{ij} c_{ik} - c_{ij} a_{ik}) = 0, \quad \sum_i (b_{ij} d_{ik} - d_{ij} b_{ik}) = 0$$

$$\sum_i (a_{ij} d_{ik} - c_{ij} b_{ik}) = \delta_{jk}.$$

The group L/C is isomorphic to the group of linear transformations (non-homogeneous) of the space (p, q) leaving the form $\sum dp_j dq_j$ invariant. T is an invariant subgroup of L and L/T is isomorphic to the group of linear and homogeneous transformations of the space (p, q) leaving $\sum dp_j dq_j$ invariant, that is, to the real symplectic group with $2n$ variables (see for example Chevalley [5] p. 23).

Subgroup Q.

Let us consider the bijective transformations
$$q'_j = q'_j(q_1, \ldots q_n), \qquad (j = 1, \ldots n)$$
of the space (q) of variables $q_1, \ldots q_n$ into itself, which are C^∞ and with Jacobian everywhere different from zero. The equations
$$s' = s, \quad p'_j = \sum p_k \frac{\partial q_k}{\partial q'_j}, \quad q'_j = q'_j(q)$$
put these transformations in correspondence with those in Γ, and extend the aforementioned transformations of the space (q) to form a subgroup Q of Γ. The transformations in Q satisfy the condition
$$\sum p'_j dq'_j = \sum p_j dq_j$$
Thus, these are homogeneous canonical transformations (Carathéodory [3] p. 97). The groups C, T and L are finite-dimensional continuous Lie groups. They depend on 1, $2n+1$ and $2n^2+3n+1$ parameters, respectively, and are locally compact. Q is an infinite-dimensional continuous group.

5   Infinitesimal Transformations in the Group Γ

Let us consider a one-parameter subgroup of the group Γ, and let us suppose that its transformations are C^∞ in the parameter. If the parameter is denoted by τ, the infinitesimal transformation of the subgroup have equations of the form
$$\frac{\delta s}{\delta \tau} = S, \quad \frac{\delta p_j}{\delta \tau} = P_j, \quad \frac{\delta q_j}{\delta \tau} = Q_j, \quad (j = 1, \ldots n).$$
S, P_j, Q_j are C^∞ functions of the variables $s, p_1, \ldots p_n, q_1, \ldots q_n$. The invariance of the differential form $\sum dp_j\, dq_j$ gives
$$\frac{\delta}{\delta \tau}\left(\sum dp_j\, dq_j\right) = \sum (dP_j\, dq_j + dp_j\, dQ_j) = d\left[\sum (P_j\, dq_j - Q_j\, dp_j)\right] = 0.$$
Thus, there exists a real function $f(p_1, \ldots p_n, q_1, \ldots q_n)$ of class C^∞ on the Euclidean space (p, q) such that
$$df = \sum (P_j\, dq_j - Q_j\, dp_j) \quad \text{or} \quad P_j = \frac{\partial f}{\partial q_j}, \quad Q_j = -\frac{\partial f}{\partial p_j}.$$
This function is only defined up to an additive constant. Let us now take into account the invariance of the form $ds - \sum p_j dq_j$. We obtain
$$\frac{\delta}{\delta \tau}\left(ds - \sum p_j dq_j\right) = dS - \sum (P_j\, dq_j + p_j dQ_j)$$

$$= d(S - \sum p_j Q_j) + \sum (Q_j dp_j - P_j dq_j)$$
$$= d\left(S + \sum p_j \frac{\partial f}{\partial p_j} - f\right) = 0.$$

The additive constant in the definition of f which has not yet been fixed will now be determined by

$$S = f - \sum p_j \frac{\partial f}{\partial p_j}.$$

Thus, a real function $f(p_1, \ldots p_n, q_1, \ldots q_n)$ of class C^∞ on the space (p, q) is uniquely associated to an infinitesimal transformation, the equations of the infinitesimal transformation being

$$\frac{\delta s}{\delta \tau} = f - \sum p_j \frac{\partial f}{\partial p_j}, \quad \frac{\delta p_j}{\delta \tau} = \frac{\partial f}{\partial q_j}, \quad \frac{\delta q_j}{\delta \tau} = -\frac{\partial f}{\partial p_j}, (j = 1, \ldots n). \quad (5.1)$$

We note that the variable s does not appear in the last two equations, and that these equations are of the same form as the canonical Hamilton equations.

Conversely, take a C^∞ function $f(p, q)$ on the space (p, q) and let us consider the differential system (5.1). To solve it, it suffices to consider the last $2n$ equations; the first one can be integrated by quadrature. Taking D to be a compact set in the space (p, q), the last $2n$ equations admit the solution

$$p_j = p_j(\tau, p^0, q^0), \quad q_j = q_j(\tau, p^0, q^0) \quad (5.2)$$

for points p_j^0, q_j^0 in D with initial conditions

$$p_j^0 = p_j(0, p^0, q^0), \quad q_j^0 = q_j(0, p^0, q^0).$$

This solution exists for values of τ in the range $\tau_1(D) \leq \tau \leq \tau_2(D)$, where $-\tau_1(D)$ and $\tau_2(D)$ are positive numbers depending on D. For these values of τ, we obtain by quadrature the value of the variable s, and the transformation obtained in this way for some part of the space (s, p, q) leaves invariant the form $ds - \sum p_j dq_j$. In order that it be possible to extend this transformation to the entire space (s, p, q), thus placing it in Γ, there must exist a positive number ϵ such that $\tau_1(D) < -\epsilon$, $\tau_2(D) > \epsilon$ for all compact sets D of the space (p, q). In this case, and only in this case, the equations (5.1) have all their integral curves defined on the interval $-\infty < \tau < +\infty$ and they generate a one-parameter subgroup of Γ. If there is no such ϵ, the group generated by (5.1) contains transformations that do not map the entire space (p, q) onto itself, but sends a portion of it to infinity: indeed, certain functions (5.2) become infinite for finite values of all their arguments.[t20]

Let us illustrate this circumstance by the following example. We assume $n = 1$ and choose $f = -pq^m$ with m integer, $m \geq 2$. Integrating the last equation of (5.1) then gives the expression

$$q = q_0(1 - (m-1)q_0^{m-1}\tau)^{\frac{1}{1-m}}$$

which becomes infinite for $(m-1)q_0^{m-1}\tau = 1$. The situation is analogous for $f = pe^q$.

The real functions $f(p,q)$ of class C^∞ which do not give rise to the above-mentioned singularities form a family $\mathcal{F}_\Gamma$; for each such function, integrating (5.1) yields a one-parameter subgroup of Γ, of class C^∞ in the parameter. *The equations (5.1) thus define a bijective correspondence between the elements of $\mathcal{F}_\Gamma$ and the C^∞ infinitesimal transformations in Γ.* For $f \in \mathcal{F}_\Gamma$, we denote by $X[f]$ the corresponding infinitesimal transformation in Γ.

Let $\phi(s, p_1, \ldots p_n, q_1, \ldots q_n)$ be a C^∞ function on the space (s, p, q). For $f \in \mathcal{F}_\Gamma$, the application of the infinitesimal transformation $X[f]$ in Γ gives

$$X[f]\phi = \left(f - \sum p_j \frac{\partial f}{\partial p_j}\right)\frac{\partial \phi}{\partial s} + (f, \phi) \tag{5.3}$$

where (f, ϕ) denotes the Poisson bracket

$$(f, \phi) = \sum \left(\frac{\partial f}{\partial q_j}\frac{\partial \phi}{\partial p_j} - \frac{\partial f}{\partial p_j}\frac{\partial \phi}{\partial q_j}\right).$$

For $f_1, f_2 \in \mathcal{F}_\Gamma$, the Lie bracket $(X[f_1], X[f_2])$ of infinitesimal transformations $X[f_1], X[f_2]$ can be defined by the relation

$$(X[f_1], X[f_2])\phi = \{X[f_1]X[f_2] - X[f_2]X[f_1]\}\phi$$

which must be valid for all C^∞ functions on the space (s, p, q). Using (5.3) and the Jacobi identity of the Poisson bracket

$$(f_1, (f_2, \phi)) + (f_2, (\phi, f_1)) + (\phi, (f_1, f_2)) = 0,$$

one obtains, after an elementary calculation, the relation

$$(X[f_1], X[f_2]) = X[(f_1, f_2)]$$

with the Poisson bracket

$$(f_1, f_2) = \sum_j \left(\frac{\partial f_1}{\partial q_j}\frac{\partial f_2}{\partial p_j} - \frac{\partial f_1}{\partial p_j}\frac{\partial f_2}{\partial q_j}\right).$$

Since we restrict ourselves to C^∞ functions, the existence of this Poisson bracket is always guaranteed. However, the symbol $X[(f_1, f_2)]$, defined by

(5.3) with $f = (f_1, f_2)$, is an infinitesimal transformation in Γ only if $(f_1, f_2) \in \mathcal{F}_\Gamma$. Similarly, for $f_1, f_2 \in \mathcal{F}_\Gamma$, and a_1 and a_2 real, one finds

$$a_1 X[f_1] + a_2 X[f_2] = X[a_1 f_1 + a_2 f_2]$$

but similarly, the symbol $X[a_1 f_1 + a_2 f_2]$ defined by (5.3) with $f = a_1 f_1 + a_2 f_2$ is only an infinitesimal transformation in Γ if $a_1 f_1 + a_2 f_2 \in \mathcal{F}_\Gamma$.

We will see that these requirements are not always satisfied.[t21] The set $\mathcal{F}_\Gamma$ is in fact not easy to understand.[t22] Although it is true that for $f \in \mathcal{F}_\Gamma$ and a real one has $af \in \mathcal{F}_\Gamma$, for $f_1, f_2 \in \mathcal{F}_\Gamma$ one cannot conclude neither $f_1 + f_2 \in \mathcal{F}_\Gamma$ nor $(f_1, f_2) \in \mathcal{F}_\Gamma$. Let us show this in two examples where $n = 1$.

1° Although $p^2/2 \in \mathcal{F}_\Gamma$ and $-q^m \in \mathcal{F}_\Gamma$ for all integer m, we do not have $p^2/2 - q^m \in \mathcal{F}_\Gamma$ when $m \geq 3$. Indeed, the corresponding differential system admits the first integral $p^2/2 - q^m = h$ and one finds

$$\tau = \int_{q_0}^{q} \frac{dq}{p} = \frac{1}{\sqrt{2}} \int_{q_0}^{q} \frac{dq}{\sqrt{q^m + h}} \ .$$

Since this integral converges for $q \to +\infty$ if $m \geq 3$, the point $p_0, q_0 > 0$ is sent to infinity by the transformation with parameter[t23]

$$\tau = \int_{q_0}^{+\infty} \frac{dq}{\sqrt{q^m + h}}, \quad h = \frac{p_0^2}{2} - q_0^m \ .$$

2° Although $p^2/2 \in \mathcal{F}_\Gamma$ and $q^{m+1}/(m+1) \in \mathcal{F}_\Gamma$ for all nonnegative integers m, we already know that[t24]

$$\left(\frac{p^2}{2}, \frac{q^{m+1}}{m+1}\right) = -pq^m \notin \mathcal{F}_\Gamma \quad \text{when } m \geq 2 \ .$$

The complications we have just described evidently stem from the fact that we are interested in transformations in Γ on the entire space (s, p, q). These complications lead us to a remark of general interest concerning infinite-dimensional groups of transformations: in the study of such groups on non-compact domains, the set of infinitesimal transformations is not necessarily closed under the Lie bracket, nor under linear combination with real coefficients; therefore, in contrast with the case for finite Lie groups, this set does not consitute a Lie algebra.

Let us now return to our objective. We can summarize the above in a theorem which makes precise the algebraic structure of the set of infinitesimal transformations in Γ.

THEOREM: *The equations (5.1) define a bijective correspondence $f \to X[f]$ between the elements of $\mathcal{F}_\Gamma$ and infinitesimal transformations in Γ. Let*

f_1 and f_2, a_1 and a_2 be real. In order that $a_1 X[f_1] + a_2 X[f_2]$ be an infinitesimal transformation in Γ, it is necessary and sufficient that $a_1 f_1 + a_2 f_2 \in \mathcal{F}_\Gamma$; then one has

$$a_1 X[f_1] + a_2 X[f_2] = X[a_1 f_1 + a_2 f_2] \ .$$

In order that the Lie bracket $(X[f_1], X[f_2])$ be an infinitesimal transformation in Γ, it is necessary and sufficient that the Poisson bracket (f_1, f_2) is in $\mathcal{F}_\Gamma$; then one has

$$(X[f_1], X[f_2]) = X[(f_1, f_2)] \ .$$

Let us also point out which elements of $\mathcal{F}_\Gamma$ correspond to infinitesimal transformations of various subgroups of Γ given in section 4. They are:

for the center C, functions $f =$ real constant;

for the subgroup T, polynomials in $p_1, \ldots p_n$, $q_1 \ldots q_n$ of degree 1 and 0, with real coefficients;

for the subgroup L, polynomials in $p_1, \ldots p_n$, $q_1 \ldots q_n$ of degree 2, 1 and 0, with real coefficients;

for the subgroup Q, functions $f(p,q) \in \mathcal{F}_\Gamma$ linear and homogeneous in $p_1, \ldots p_n$.

6 A Lie Algebra of Infinitesimal Transformations in Γ

One can easily define, in the heart of $\mathcal{F}_\Gamma$, subsets that are closed under linear combination with real coefficients and under Poisson brackets. Such subsets constitute Lie algebras, and the same reasoning holds for sets corresponding to infinitesimal transformations in Γ.

Let us call Γ the set of real C^∞ functions $f(p_1, \ldots p_n, q_1, \ldots q_n)$ on the space (p, q), enjoying the following properties:

(a) With the notation $\rho^2 = \sum(p_j^2 + q_j^2)$, we have $f(p,q) = \mathcal{O}(\rho^2)$ for $\rho \to +\infty$. [f]

(b) For $m = 1, 2, \ldots$, all partial derivatives of order m of f satisfy the condition $f^{(m)}(p,q) = \mathcal{O}(\rho^{2-m})$ for $\rho \to +\infty$.

---

[f] We say that $g(p,q) = \mathcal{O}(\rho^m)$ for $\rho \to +\infty$ when $\rho^{-m} g(p,q)$ remains bounded for sufficiently large values of ρ.

In stating these properties one could replace ρ by the square root of any positive-definite quadratic form in p_j, q_j.[t25]

Given $f_1, f_2 \in \mathcal{F}$, one immediately sees that $(f_1, f_2) \in \mathcal{F}$ and that $a_1 f_1 + a_2 f_2 \in \mathcal{F}$ for all real values of a_1, a_2. For the composition law defined by the Poisson bracket, the set $\mathcal{F}$ thus constitutes a Lie algebra over the field of real numbers.

We will now show that $\mathcal{F}$ is a subgroup of $\mathcal{F}_\Gamma$. It is enough to show that for $f \in \mathcal{F}$ all integral curves of the differential system

$$\frac{dp_j}{d\tau} = \frac{\partial f}{\partial q_j}, \quad \frac{dq_j}{d\tau} = -\frac{\partial f}{\partial p_j}$$

are defined on the interval $-\infty < \tau < +\infty$.

Consider an integral curve and suppose that $\tau' < \tau < \tau''$ is the maximal interval for which it exists. One can assume $\tau' < 0 < \tau''$. Let us first suppose that for $\tau \to \tau''$, the curve stays within a compact set D. If the curve is defined for a value τ_0 of the parameter τ, using (5.2) one can define it on the interval

$$\tau_0 + \tau_1(D) \leq \tau \leq \tau_0 + \tau_2(D).$$

Therefore $\tau'' = +\infty$. If on the other hand the curve does not stay within any compact set for $\tau \to \tau''$, call e the set of values of τ for which[t26]

$$0 < \tau < \tau'', \quad \frac{d\rho}{d\tau} \geq 0 \quad \Longleftrightarrow \quad \tau \in e.$$

We have

$$\tau'' = \int_0^{\tau''} d\tau \geq \int_e \tau = \int_e \frac{d\rho}{d\tau} d\rho.$$

But

$$\frac{d\rho}{d\tau} = \sum_j \left(\frac{p_j}{\rho} \frac{\partial f}{\partial q_j} - \frac{q_j}{\rho} \frac{\partial f}{\partial p_j} \right);$$

according to (b), for ρ sufficiently large, we now have

$$\left| \frac{d\rho}{d\tau} \right| \leq A\rho$$

where A is a positive number. When τ varies in e, ρ takes values in a set which contains the interval $\rho_0 \leq \rho < +\infty$ for ρ_0 sufficiently large. Thus

$$\tau'' \geq \int_{\rho_0}^{+\infty} \left(\frac{d\rho}{d\tau} \right)^{-1} d\rho \geq \frac{1}{A} \int_{\rho_0}^{+\infty} \frac{d\rho}{\rho}.$$

The last integral being divergent, we have again $\tau'' = +\infty$. Similarly we find $\tau' = -\infty$.

Since $\mathcal{F} \in \mathcal{F}_\Gamma$, infinitesimal transformations of the group Γ correspond to elements f of $\mathcal{F}$. Denote by $X[\mathcal{F}]$ the set of such transformations. By the theorem of the preceding section, $X[\mathcal{F}]$ *constitutes a Lie algebra over the field of real numbers with composition law defined by the Lie bracket, and the correspondence $f \to X[f]$ is an isomorphism between the Lie algebras $\mathcal{F}$ and $X[\mathcal{F}]$.* We remark that $X[\mathcal{F}]$ contains the infinitesimal transformations of the subgroups C, T, L of section 4.

CHAPTER III

Unitary Representations $\mathcal{R}$ and $\mathcal{R}^{(\alpha)}$ of the Group Γ.

In a separable Hilbert space, we construct a unitary representation $\mathcal{R}$ for the group Γ, and a family of unitary representations $\mathcal{R}^{(\alpha)}$ depending on the real parameter α. We show that the representation $\mathcal{R}$ is reducible to the continuous sum of representations $\mathcal{R}^{(\alpha)}$.

7 The Representation $\mathcal{R}$

As we have seen in section 3, transformations in Γ leave invariant the volume elements $ds\, dp_1, \ldots dp_n\, dq_1 \ldots dq_n$ and $dp_1 \ldots dp_n dq_1 \ldots dq_n$ of the spaces (s, p, q) and (p, q). Using this remark and applying a well-known method, we are going to construct certain unitary representations of the group Γ in a separable Hilbert space.

Let us use the notation Ω for the points of the space (s, p, q) of the variables $s, p_1, \ldots p_n, q_1 \ldots q_n$ and the notation $d\Omega$ for the volume element $ds\, dp_1 \ldots dp_n\, dq_1 \ldots dq_n$. As we know, the elements of a separable Hilbert space $\mathcal{H}_{2n+1}$ are complex-valued functions $\phi(\Omega)$, $\Psi(\Omega), \ldots$, measurable and square integrable on the space (s, p, q), and with scalar product

$$(\phi, \Psi) = \int \bar{\phi}(\Omega) \Psi(\Omega)\, d\Omega \tag{7.1}$$

integrated over the entire space (s, p, q). The function $\bar{\phi}(\Omega)$ is the complex conjugate of $\phi(\Omega)$. Let γ be a transformation in the group Γ. The transformation of the point Ω under the inverse transformation γ^{-1} will be denoted by $\gamma^{-1}\Omega$. For every element $\phi(\Omega)$ of $\mathcal{H}_{2n+1}$ we define the function $U_\gamma \phi(\Omega)$ by the equation

$$U_\gamma \phi(\Omega) = \phi(\gamma^{-1}\Omega)\,.$$

The operation U_γ satisfies for $\phi, \Psi \in \mathcal{H}_{2n+1}$

$$(U_\gamma \phi, U_\gamma \Psi) = \int \bar{\phi}(\gamma^{-1}\Omega) \Psi(\gamma^{-1}\Omega)\, d\Omega = \int \bar{\phi}(\Omega) \Psi(\Omega)\, d(\gamma\Omega) = (\phi, \Psi)\,,$$

due to the invariance of the volume element: $d(\gamma\Omega) = d\Omega$. We see that U_γ is a linear transformation of the space $\mathcal{H}_{2n+1}$ into itself, leaving the scalar product

(7.1) invariant. It is therefore a unitary transformation in $\mathcal{H}_{2n+1}$ (Stone [30] p.76). For $\gamma_1, \gamma_2 \in \Gamma$, we have

$$U_{\gamma_1\gamma_2}\phi(\Omega) = \phi(\gamma_2^{-1}\gamma_1^{-1}\Omega) = U_{\gamma_1}\phi(\gamma_2^{-1}\Omega) = U_{\gamma_1}U_{\gamma_2}\phi(\Omega) .$$

Thus $U_{\gamma_1\gamma_2} = U_{\gamma_1}U_{\gamma_2}$. For the unit element γ_0 of the group Γ, U_{γ_0} is the identity transformation of $\mathcal{H}_{2n+1}$. It follows that *the map $\gamma \to U_\gamma$ is a representation of the group Γ by unitary transformations of the Hilbert space $\mathcal{H}_{2n+1}$*. We denote this representation by $\mathcal{R}$. We immediately see that if the elements γ_1 and γ_2 of Γ are different, the unitary transformations $U_{\gamma_1}, U_{\gamma_2}$ are also different.[t27] *Thus, the representation $\mathcal{R}$ is faithful.*

8 The Representations $\mathcal{R}^{(\alpha)}$

Now let us consider the space (p, q) of variables $p_1, \ldots p_n, q_1, \ldots q_n$. We will use ω for points in this space and the notation $d\omega$ for its volume element $dp_1 \ldots dp_n \, dq_1 \ldots dq_n$. We introduce the separable Hilbert space $\mathcal{H}_{2n}$ of complex-valued functions $\varphi(\omega)$, $\psi(\omega)$, ..., measurable and square integrable on the space (s, p, q) and with scalar product

$$(\varphi, \psi) = \int \bar{\varphi}(\omega)\psi(\omega) \, d\omega \tag{8.1}$$

integrated over the entire space (p, q).

Now, let γ be a transformation in the group Γ represented by equations (3.8), (3.9). We denote by ω the point p_j, q_j, by ω' the point p'_j, q'_j and we write the transformation in the form

$$\omega' = \gamma\omega , \quad s' = s + \pi_\gamma(\omega)$$

Taking α to be a real number, and $\varphi(\omega)$ an element of the Hilbert space $\mathcal{H}_{2n}$, we consider the function

$$U_\gamma^{(\alpha)}\varphi(\omega) = e^{i\alpha\pi_\gamma(\gamma^{-1}\omega)}\varphi(\gamma^{-1}\omega) .$$

Once more, we have for $\varphi(\omega), \psi(\omega) \in \mathcal{H}_{2n}$

$$(U_\gamma^{(\alpha)}\varphi, U_\gamma^{(\alpha)}\psi) = (\varphi, \psi)$$

following from the invariance of the volume element: $d(\gamma\omega) = d\omega$. The operation $U_\gamma^{(\alpha)}$ is therefore a unitary transformation in $\mathcal{H}_{2n}$. For $\gamma_1, \gamma_2 \in \Gamma$, we have

$$\pi_{\gamma_1\gamma_2}(\omega) = \pi_{\gamma_2}(\omega) + \pi_{\gamma_1}(\gamma_2\omega) .$$

We easily find
$$U^{(\alpha)}_{\gamma_1\gamma_2} = U^{(\alpha)}_{\gamma_1} U^{(\alpha)}_{\gamma_2} .$$

Further, for the unit element γ_0 of Γ, $U^{(\alpha)}_{\gamma_0}$ reduces to the identity transformation of $\mathcal{H}_{2n}$. Thus, *for every real number α, the map $\gamma \to U^{(\alpha)}_\gamma$ is a representation of the group Γ by unitary transformations of the Hilbert space $\mathcal{H}_{2n}$*. We denote this representation by $\mathcal{R}^{(\alpha)}$. We easily check that $\mathcal{R}^{(\alpha)}$ is a faithful representation of Γ for $\alpha \neq 0$ and that $\mathcal{R}^{(0)}$ is a faithful representation of the quotient group Γ/C.

The representation $\mathcal{R}^{(0)}$ has been used in ergodic theory, by Koopman and von Neumann, to represent the time evolution of a conservative dynamical system by a one-parameter group of transformations (see notably Koopman [15] and von Neumann [24]).

9 Reduction of the Representation $\mathcal{R}$

Consider the Hilbert space $\mathcal{H}'_{2n+1}$ of complex-valued functions $\varphi'(\alpha, \omega)$ that depend on the real variable α, $(-\infty < \alpha < +\infty)$ and on the point ω in the space (p, q); considered as functions of $2n + 1$ variables $\alpha, p_1, \ldots p_n, q_1, \ldots q_n$, the functions $\varphi'(\alpha, \omega)$ are measurable and square integrable. The scalar product in $\mathcal{H}'_{2n+1}$ is defined by the integral

$$(\varphi', \psi') = \int \bar{\varphi}'(\alpha, \omega) \psi'(\alpha, \omega) \, d\alpha \, d\omega \qquad (9.1)$$

over the entire space (α, p, q) of variables $\alpha, p_1, \ldots p_n, q_1, \ldots q_n$. Take φ' to be an element of $\mathcal{H}'_{2n+1}$ defined by the function $\varphi'(\alpha, \omega)$. According to the Fubini theorem (see for example La Vallée Poussin[16] p. 57), for almost all values of α (that is, with the exception of values of α belonging to a set of measure zero) the function $\varphi'(\alpha, \omega)$ is a measurable and square integrable function of the point ω on the space (p, q) and thus defines an element of the space $\mathcal{H}_{2n}$ that we call $\varphi^{(\alpha)}$. Consequently, given two elements φ', ψ' of $\mathcal{H}'_{2n+1}$, for almost all values of α the functions $\varphi'(\alpha, \omega)$ and $\psi'(\alpha, \omega)$ are simultaneously measurable and square integrable and therefore define the elements $\varphi^{(\alpha)}$, $\psi^{(\alpha)}$ of $\mathcal{H}_{2n}$. Furthermore, we have

$$(\varphi', \psi') = \int_{-\infty}^{+\infty} d\alpha \int \bar{\varphi}'(\alpha, \omega) \psi'(\alpha, \omega) \, d\omega = \int_{\infty}^{+\infty} (\varphi^{(\alpha)}, \psi^{(\alpha)}) \, d\alpha \qquad (9.2)$$

where $(\varphi^{(\alpha)}, \psi^{(\alpha)})$ denotes the scalar product (8.1) of $\mathcal{H}_{2n}$.

For every value of α, take a copy $\mathcal{H}^{(\alpha)}_{2n}$ of the space $\mathcal{H}_{2n}$. We see that in every $\mathcal{H}^{(\alpha)}_{2n}$, except those for which α is in a set of measure zero, an element

φ' of $\mathcal{H}'_{2n+1}$ corresponds to an element $\varphi^{(\alpha)}$ that we may call its component in $\mathcal{H}_{2n}^{(\alpha)}$.[28] With this in mind, and using the formula (9.2), one can say that $\mathcal{H}'_{2n+1}$ is the *continuous direct sum* of spaces $\mathcal{H}_{2n}^{(\alpha)}$ and that we can write the relation between φ' and its components as $\varphi' = \int_\oplus \varphi^{(\alpha)}$. (For this notation see Mautner [18]). However, contrary to the case of ordinary direct sums in Hilbert space (direct sums of a finite number of terms), the components of an element of $\mathcal{H}'_{2n+1}$ cannot depend on α in an arbitrary manner: $\varphi^{(\alpha)}(\omega)$ must be measurable and square integrable as a function of the variables $\alpha, p_1, \ldots, p_n, q_1, \ldots, q_n$. Here, we encounter a rather special case of the notion of generalized direct sums of Hilbert spaces, introduced by von Neumann (von Neumann[25]; also see Godement [11] for an extension to Banach spaces). Now, let us denote by s_Ω and ω_Ω the coordinate s and the set of coordinates $p_1, \ldots p_n, q_1, \ldots q_n$ of a point Ω on the space (s, p, q). It is well known that the Fourier transform

$$\phi(\Omega) = \frac{1}{\sqrt{2\pi}} \int_{-\infty}^{+\infty} \exp(-i\alpha s_\Omega) \cdot \varphi'(\alpha, \omega_\Omega) \, d\alpha \qquad (9.3)$$

defines a bijective and linear map F of the space $\mathcal{H}'_{2n+1}$ onto the space $\mathcal{H}_{2n+1}$, which transforms the scalar product of $\mathcal{H}'_{2n+1}$, equation (9.2), into the scalar product of $\mathcal{H}_{2n+1}$, equation (7.1). The inverse map F^{-1} is given by

$$\varphi'(\alpha, \omega) = \frac{1}{\sqrt{2\pi}} \int_{-\infty}^{+\infty} \exp(i\alpha s_\Omega) \cdot \phi(\Omega) \, ds_\Omega \qquad (9.4)$$

where the integrals are to be performed at fixed $\omega = \omega_\Omega$. The integrals (9.3) and (9.4) are convergent with respect to the norms of $\mathcal{H}_{2n+1}$ and $\mathcal{H}'_{2n+1}$, respectively.

Let us move on to the representation $\mathcal{R}$ and take an element γ of the group Γ; apply the transformation U_γ to the element (9.3) of $\mathcal{H}_{2n+1}$. Noting that

$$s_{\gamma^{-1}\Omega} = s_\Omega - \pi_\gamma(\gamma^{-1}\omega_\Omega) \, ,$$

we find by an easy calculation

$$U_\gamma \phi(\Omega) = \frac{1}{\sqrt{2\pi}} \int_{-\infty}^{+\infty} \exp(i\alpha s_\Omega) \cdot \exp[i\alpha \pi_\gamma(\gamma^{-1}\omega_\Omega)] \cdot \varphi'(\alpha, \gamma^{-1}\omega_\Omega) \, d\alpha \, .$$

Therefore, the unitary transformation $F^{-1} U_\gamma F = V_\gamma$ in the space $\mathcal{H}'_{2n+1}$ is the following:

$$V_\gamma \varphi'(\alpha, \omega) = \exp[i\alpha \pi_\gamma(\gamma^{-1}\omega)] \cdot \varphi'(\alpha, \gamma^{-1}\omega) \, . \qquad (9.5)$$

For almost all values of α, the element φ' of $\mathcal{H}'_{2n+1}$ defined by the function $\varphi'(\alpha,\omega)$ admits components $\varphi^{(\alpha)}$. According to (9.5), for the same values of α, $V_\gamma \varphi'$ admits components $U_\gamma^{(\alpha)}\varphi^{(\alpha)}$ where the transformation $U_\gamma^{(\alpha)}$ is in the representation $\mathcal{R}^{(\alpha)}$ (section 8). Thus, in the above notation we have

$$V_\gamma \int_\oplus \varphi^{(\alpha)} = F^{-1} U_\gamma F \int_\oplus \varphi^{(\alpha)} = \int_\oplus U_\gamma^{(\alpha)}\varphi^{(\alpha)} . \tag{9.6}$$

The map $\gamma \to V_\gamma$ is a representation of the group Γ equivalent to $\mathcal{R}$. Thus we can say that *the representation $\mathcal{R}$ admits the reduction (9.6) into the continuous family of representations $\mathcal{R}^{(\alpha)}$, $(-\infty < \alpha < +\infty)$*, the precise meaning of this statement being given in the previous paragraphs.

10 Linear Manifolds Invariant Under the Representation $\mathcal{R}$

In sections 18 and 19, we will establish that the representations $\mathcal{R}^{(\alpha)}$ are irreducible. Since they relate elements of the center of Γ to different multiples of the identity transformation, they are pairwise inequivalent. It is then possible to deduce which are the linearly closed[t29] manifolds of the space $\mathcal{H}_{2n+1}$ left invariant by the representation $\mathcal{R}$.

Let $\mathcal{E}_\alpha$ be a measurable set with nonzero measure (finite or infinite) on the real line $-\infty < \alpha < +\infty$. In $\mathcal{H}_{2n+1}$, consider functions $\varphi'(\alpha,\omega)$ for which $\alpha \in \mathcal{E}_\alpha$, $\omega \in$ space (p,q). These functions form a linearly closed manifold in $\mathcal{H}'_{2n+1}$ that we shall denote by $\mathcal{M}'(\mathcal{E}_\alpha)$. The linearly closed manifolds $\mathcal{M}'(\mathcal{E}_\alpha)$, $\mathcal{M}'(\mathcal{E}'_\alpha)$ coincide when the sets $\mathcal{E}_\alpha$, $\mathcal{E}'_\alpha$ differ only by a set of measure zero, and only in that case. We call $\mathcal{M}(\mathcal{E}_\alpha)$ the image of $\mathcal{M}'(\mathcal{E}_\alpha)$ in $\mathcal{H}_{2n+1}$ under the map F. Thus, we have the following proposition:

THEOREM: *The only linearly closed manifolds of $\mathcal{H}_{2n+1}$ that are invariant for the representation $\mathcal{R}$ of the group Γ are the manifolds where $\mathcal{E}_\alpha$ is any measurable set of nonzero measure on the real line $-\infty < \alpha < +\infty$.*

This result can easily be derived from von Neumann's theory on reduction of the ring of operators in a Hilbert space (von Neumann [25]; we will base our discussion on lemma 13, p. 459). However, for our purposes, we can find this result without recourse to the general theory. This is what we shall show presently. It is clearly enough to prove that the only linearly closed manifolds of $\mathcal{H}'_{2n+1}$ that are invariant under the transformations V_γ, $(\gamma \in \Gamma)$ are the manifolds $\mathcal{M}'(\mathcal{E}_\alpha)$ defined above.

When we prove irreducibility of the representations $\mathcal{R}^{(\alpha)}$ (see section 18), we will see that there exists a countable series $\gamma_1, \ldots \gamma_k, \ldots$ of elements of the group Γ such that for every $\alpha \neq 0$, the Hilbert space $\mathcal{H}_{2n}^{(\alpha)}$ is irreducible under the unitary transformations $U_{\gamma_1}^{(\alpha)}, \ldots U_{\gamma_k}^{(\alpha)}, \ldots$ [t30] Let $\mathcal{N}'$ be a linearly closed manifold in $\mathcal{H}'_{2n+1}$, invariant under unitary transformations $V_\gamma = F^{-1} U_\gamma F$, ($\gamma \in \Gamma$). We denote by $\mathcal{P}'$ the linearly closed manifold of elements in $\mathcal{H}'_{2n+1}$ orthogonal to $\mathcal{N}'$. This manifold is also invariant under the V_γ.

When a linearly closed manifold $\mathcal{V}'$ of $\mathcal{H}'_{2n+1}$ is invariant under the V_γ, it is invariant, in particular, under the one-parameter group of unitary transformations that arises when γ varies in the center C of the group Γ. To this one-parameter subgroup corresponds a resolution of the identity which, by (9.5) for $\gamma \in C$, is composed of projection operators $E_c(\lambda)$ on the manifolds $\mathcal{M}'(I_\gamma)$ where I_γ is the interval $-\infty < \alpha \leq \lambda$. Since $\mathcal{V}'$ is invariant under V_γ, ($\gamma \in C$), it is invariant under the operators $E_c(\lambda)$ (see Stone [31]). This holds in particular for the manifolds $\mathcal{N}'$ and $\mathcal{P}'$.

Now let
$$\varphi' \in \mathcal{N}', \quad \psi' \in \mathcal{P}'$$
and
$$\varphi' = \int_\oplus \varphi^{(\alpha)}, \quad \psi' = \int_\oplus \psi^{(\alpha)}.$$

Using (9.2), the orthogonality of the manifolds $\mathcal{N}'$, $\mathcal{P}'$, and their invariance under V_γ, one has
$$(\psi', V_\gamma \varphi') = \int_{-\infty}^{+\infty} (\psi^{(\alpha)}, U_\gamma^{(\alpha)} \varphi^{(\alpha)}) \, d\alpha = 0.$$

This relation remains true when one replaces ψ' by $E_c(\lambda)\psi'$, which amounts to annihilating the components $\psi^{(\alpha)}$ for $\alpha > \lambda$. We thus have for all λ
$$(E_c(\lambda)\psi', V_\gamma \varphi') = \int_{-\infty}^{\lambda} (\psi^{(\alpha)}, U_\gamma^{(\alpha)} \varphi^{(\alpha)}) \, d\alpha = 0.$$

This implies
$$(\psi^{(\alpha)}, U_\gamma^{(\alpha)} \varphi^{(\alpha)}) = 0$$
for all values of α, except for those in a set of measure zero $e(\gamma)$ which, for given φ' and ψ', depends only on γ.

Now consider the subgroup Δ of Γ of products of a finite number of transformations of the series $\gamma_1, \ldots \gamma_k, \ldots$. The subgroup Δ contains a countable infinity of elements; the set $e_\Delta = \bigcup_{\gamma \in \Delta} e(\gamma)$ is thus of measure zero. For every α not in e_Δ one also has
$$(\psi^{(\alpha)}, U_\gamma^{(\alpha)} \varphi^{(\alpha)}) = 0 \tag{10.1}$$
for all $\gamma \in \Delta$. The linearly closed manifold generated in $\mathcal{H}_{2n}$ by the vectors $U_\gamma^{(\alpha)} \varphi^{(\alpha)}$, ($\gamma \in \Delta$) is invariant under the $U_{\gamma_k}^{(\alpha)}$. Thus, for $\alpha \neq 0$, it reduces

either to the zero vector $\varphi^{(\alpha)} = 0$, or to the space $\mathcal{H}_{2n}$ and therefore, by (10.1), to $\psi^{(\alpha)} = 0$. Denote by $\mathcal{E}_{\alpha,\varphi'}$ the set of values of α for which $\varphi^{(\alpha)}$ is different from zero; it is determined up to a set of measure zero. The linearly closed manifold $\mathcal{M}'(\mathcal{E}_{\alpha,\varphi'})$ is orthogonal to ψ'. Thus, this manifold is orthogonal to $\mathcal{P}'$ and contained in $\mathcal{N}'$. Take $\mathcal{E}_{\alpha,\mathcal{N}'}$ to be the union of sets $\mathcal{E}_{\alpha,\varphi'}$ when φ' describes a countable basis of $\mathcal{N}'$. The manifold $\mathcal{M}'(\mathcal{E}_{\alpha,\mathcal{N}'})$ does not depend on the choice of basis. It is contained within $\mathcal{N}'$; at the same time, it contains every vector of a basis of $\mathcal{N}'$. Therefore we have $\mathcal{N}' = \mathcal{M}'(\mathcal{E}_{\alpha,\mathcal{N}'})$, which proves the proposition.

CHAPTER IV

Infinitesimal Transformations in the Representations $\mathcal{R}$ and $\mathcal{R}^{(\alpha)}$.

To infinitesimal transformations of the group Γ correspond, in each of the representations $\mathcal{R}$, $\mathcal{R}^{(\alpha)}$, hypermaximally antisymmetric operators $-iH[f]$, $-iH^{(\alpha)}$. In the Hilbert space of these representations, one can identify an invariant domain in which these operators are essentially hypermaximally symmetric. It follows that every Lie algebra of infinitesimal transformations in Γ corresponds, in the representations $\mathcal{R}$, $\mathcal{R}^{(\alpha)}$, to Lie algebras of operators in the sense of Segal.

11 Infinitesimal Transformations in the Representations $\mathcal{R}$

To each real C^∞ function $f(p_1, \ldots p_n, q_1 \ldots q_n)$ on the space (p, q), belonging to a family $\mathcal{F}_\Gamma$ defined in section 5, there corresponds an infinitesimal transformation $X[f]$ of the group Γ. The infinitesimal transformation $X[f]$ generates a continuous one-parameter subgroup in Γ, with parameter τ, and elements labelled $\gamma_\tau^{(f)}$. As usual, Ω, Ω' denote points of the space (s, p, q). Suppose that

$$\Omega' = \gamma_\tau^{(f)} \Omega \,.$$

Thus, the $2n+1$ coordinates of Ω' are C^∞ functions with parameter τ of the $2n+1$ coordinates of Ω; these functions are obtained by integrating the differential system (5.1). Now consider the representation $\mathcal{R}$ of the one-parameter group of unitary transformations $U_{\gamma_\tau^{(f)}}$ in the space $\mathcal{H}_{2n+1}$. Applying a well-known method of proof (see for example Weil [33], p. 40), we will show that the transformation $U_{\gamma_\tau^{(f)}}$ is strongly continuous in τ. Let ϕ be an element of $\mathcal{H}_{2n+1}$, and η an arbitrary positive number. We can find a function $\Psi(\Omega)$ continuous on the space (s, p, q), zero outside a compact set, such that

$$\|\phi - \Psi\| < \frac{\eta}{3} \,.$$

We are using the standard notation for the norm in Hilbert space. The U_γ being unitary, we will have for all σ, τ

$$\|U_{\gamma_\tau^{(f)}} \phi - U_{\gamma_\tau^{(f)}} \Psi\| < \frac{\eta}{3}, \quad \|U_{\gamma_\sigma^{(f)}} \phi - U_{\gamma_\sigma^{(f)}} \Psi\| < \frac{\eta}{3} \,.$$

But $U_{\gamma_\sigma^{(f)}}\Psi$ is a continuous function on the space (s,p,q) depending continuously on the parameter τ; when τ remains bounded, this function vanishes outside a fixed compact set of the space (s,p,q). For σ sufficiently close to τ, we will therefore have

$$||U_{\gamma_\tau^{(f)}}\Psi - U_{\gamma_\sigma^{(f)}}\Psi|| < \frac{\eta}{3},$$

hence

$$||U_{\gamma_\tau^{(f)}}\phi - U_{\gamma_\sigma^{(f)}}\phi|| < \eta.$$

Thus, the proposition is proven. Moreover, note that it would have been sufficient to show that for $\phi, \phi' \in \mathcal{H}_{2n+1}$, the scalar product $(\phi, U_{\gamma_\tau^{(f)}}\phi')$ is a measurable function of ϕ. Strong continuity then follows from the theorem of Stone (section 2, theorems A and B).

Since

$$U_{\gamma_\sigma^{(f)}} U_{\gamma_\tau^{(f)}} = U_{\gamma_{\sigma+\tau}^{(f)}}$$

we can apply the theorem of Stone (section 2, theorems B and C). There exists a hypermaximally symmetric operator $H[f]$, with domain $D[f]$, such that

$$U_{\gamma_\tau^{(f)}} = e^{i\tau H[f]}$$

for all values of the real parameter τ. The domain $D[f]$ is the set of elements $\phi \in \mathcal{H}_{2n+1}$ for which the limit

$$\lim_{t \to 0} \tau^{-1}[U_{\gamma_\tau^{(f)}}\phi - \phi] \tag{11.1}$$

exists with respect to the norm of $\mathcal{H}_{2n+1}$. For $\phi \in D[f]$ this limit equals $iH[f]\phi$. If $\phi(\Omega)$ is a continuous function with partial derivatives measurable on the space (s,p,q), and if the function vanishes outside a compact set, the limit (11.1) exists; thus $\phi \in D[f]$ and by (5.3) we easily find

$$H[f]\phi = iX[f]\phi \tag{11.2}$$
$$= i\left(f - \sum p_j \frac{\partial f}{\partial p_j}\right)\frac{\partial \phi}{\partial s} + i\sum \left(\frac{\partial f}{\partial q_j}\frac{\partial \phi}{\partial p_j} - \frac{\partial f}{\partial p_j}\frac{\partial \phi}{\partial q_j}\right).$$

We denote by $\mathcal{D}_{2n+1}$ the set of C^∞ functions $\phi(\Omega)$ on the space (s,p,q), vanishing outside compact sets. We see that for all $f \in \mathcal{F}_\Gamma$, the set $\mathcal{D}_{2n+1}$ is contained in $D[f]$, that (11.2) is valid for all $\phi \in \mathcal{D}_{2n+1}$, and that $H[f]$ is an automorphism of $\mathcal{D}_{2n+1}$:[t31]

$$\mathcal{D}_{2n+1} \subset D[f], \quad H[f]\mathcal{D}_{2n+1} \subset \mathcal{D}_{2n+1}.$$

12 Infinitesimal Transformations in $\mathcal{R}^{(\alpha)}$. Properties of the Obtained Operators

Entirely analogous considerations apply to the representation $\mathcal{R}^{(\alpha)}$ for every real value of α. We apply the notation and formulas of section 8 to the elements $\gamma_\tau^{(f)}$ of the one-parameter group generated by the infinitesimal transformation $X[f]$ ($f \in \mathcal{F}_\Gamma$). In the equations

$$\omega' = \gamma_\tau^{(f)} \omega, \quad s' = s + \pi_{\gamma_\tau^{(f)}}(\omega),$$

the $2n$ coordinates of ω' and the function $\pi_{\gamma_\tau^{(f)}}$ are C^∞ functions of the real parameter α and of the coordinates of ω. By reasoning analogous to that of the preceding section, we show that in the Hilbert space $\mathcal{H}_{2n}$, the unitary transformation $U_{\gamma_\tau^{(f)}}^{(\alpha)}$ is strongly continuous in τ. Applying the theorem of Stone, one thus obtains for all real values of τ

$$U_{\gamma_\tau^{(f)}}^{(\alpha)} = e^{i\tau H^{(\alpha)}[f]},$$

where $H^{(\alpha)}[f]$ is a hypermaximally symmetric operator, its domain $D^{(\alpha)}[f]$ being the set of elements $\varphi \in \mathcal{H}_{2n}$ for which the limit

$$\lim_{\tau \to 0} \tau^{-1}[U_{\gamma_\tau^{(f)}}^{(\alpha)} \varphi - \varphi]$$

exists in the norm of $\mathcal{H}_{2n}$. For $\varphi \in D^{(\alpha)}[f]$ this limit equals $iH^{(\alpha)}[f]\varphi$. The domain $D^{(\alpha)}[f]$ contains the continuous functions $\varphi(\omega)$ with partial derivatives measurable on the space (p, q), vanishing outside a compact set. For such functions, an easy calculation gives

$$H^{(\alpha)}[f]\varphi = \alpha \left(f - \sum p_j \frac{\partial f}{\partial p_j} \right) \varphi + i \sum \left(\frac{\partial f}{\partial q_j} \frac{\partial \varphi}{\partial p_j} - \frac{\partial f}{\partial p_j} \frac{\partial \varphi}{\partial q_j} \right). \quad (12.1)$$

In particular, this equation holds when φ is in the set $\mathcal{D}_{2n}$ of C^∞ functions on the space (p, q), vanishing outside compact sets. Thus, for every element f of $\mathcal{F}_\Gamma$ and for every real value of α we have

$$\mathcal{D}_{2n} \subset D^{(\alpha)}[f], \quad H^{(\alpha)}[f]\mathcal{D}_{2n} \subset \mathcal{D}_{2n}. \quad (12.2)$$

We have constructed hypermaximally symmetric operators $H[f]$ in bijective correspondence with the elements f of $\mathcal{F}_\Gamma$, and for every real value of α, hypermaximally symmetric operators $H^{(\alpha)}[f]$. On the other hand, we saw in section 5 that the correspondence $f \to X[f]$ between elements of $\mathcal{F}_\Gamma$ and infinitesimal transformations of Γ enjoy the following properties:

$$X[a_1 f_1 + a_2 f_2] = X[a_1 f_1 + a_2 f_2] \quad \text{if } f_1, f_2, a_1 f_1 + a_2 f_2 \in \mathcal{F}_\Gamma \quad (12.3)$$
$$(X[f_1], X[f_2]) = X[(f_1, f_2)] \quad \text{if } f_1, f_2, (f_1, f_2) \in \mathcal{F}_\Gamma. \quad (12.4)$$

In the representation $\mathcal{R}$, $X[f]$ is represented via (11.2) by the hypermaximally antisymmetric operator $-iH[f]$. Thus, we expect the properties (12.3), (12.4) to give the following relations:

$$H[a_1 f_1 + a_2 f_2]\phi = a_1 H[f_1] + a_2 H[f_2]\phi \qquad (12.5)$$
$$\text{if } f_1, \, f_2, \, a_1 f_1 + a_2 f_2 \in \mathcal{F}_\Gamma$$
$$iH[(f_1, f_2)]\phi = H[f_1]H[f_2] - H[f_2]H[f_1]\phi \qquad (12.6)$$
$$\text{if } f_1, \, f_2, \, (f_1, f_2) \in \mathcal{F}_\Gamma \, .$$

for elements ϕ of $\mathcal{H}_{2n+1}$ whenever the operators on both sides of the equation are defined. Using the explicit expressions (11.2) for $H[f]$ one easily verifies these relations in the case where ϕ is a sufficiently differentiable function. It remains to convince oneself that the equalities (12.5) and (12.6) also hold for maximal extensions of the operators appearing in the equations. That this is true follows from the first part of the following theorem, which plays a fundamental role to be established later.

THEOREM: *For all $f \in \mathcal{F}_\Gamma$, the restriction of the operator $H[f]$ to the domain [9] $\mathcal{D}_{2n+1}$ is essentially hypermaximally symmetric; for all $f \in \mathcal{F}_\Gamma$ and all real α, the restriction of the operator $H^{(\alpha)}[f]$ to the domain $\mathcal{D}_{2n}$ is essentially hypermaximally symmetric.*

Using (11.2), the relations (12.5) and (12.6) immediately follow for elements ϕ of $\mathcal{D}_{2n+1}$. We then immediately have the following proposition which gives precise meaning to relations (12.5) and (12.6).

COROLLARY I: *Let f_1 and f_2 be two elements of $\mathcal{F}_\Gamma$, and a_1 and a_2 two real numbers. If $a_1 f_1 + a_2 f_2 \in \mathcal{F}_\Gamma$, the operator $a_1 H[f_1] + a_2 H[f_2]$ defined on $\mathcal{D}_{2n+1}$ is essentially hypermaximally symmetric and its hypermaximally symmetric extension is the operator $H[a_1 f_1 + a_2 f_2]$. If $(f_1, f_2) \in \mathcal{F}_\Gamma$, the operator $H[f_1]H[f_2] - H[f_2]H[f_1]$ defined on $\mathcal{D}_{2n+1}$ is essentially hypermaximally antisymmetric and its unique hypermaximally antisymmetric extension is the operator $iH[(f_1, f_2)]$.*

Entirely analogous considerations apply to the representations $\mathcal{R}^{(\alpha)}$. In this case, the infinitesimal transformation $X[f]$ of Γ, $(f \in \mathcal{F}_\Gamma)$, is represented by the hypermaximally antisymmetric operator $-iH^{(\alpha)}[f]$, with the following property, which can be proven from (12.1) for the elements of $\mathcal{D}_{2n}$.

COROLLARY II: *Let $f_1, f_2, \in \mathcal{F}_\Gamma$, α, a_1, a_2 real. If $a_1 f_1 + a_2 f_2 \in \mathcal{F}_\Gamma$, the operator $a_1 H^{(\alpha)}[f_1] + a_2 H^{(\alpha)}[f_2]$ defined on $\mathcal{D}_{2n}$ is essentially hypermaximally symmetric and its hypermaximally symmetric extension is the operator $H^{(\alpha)}[a_1 f_1 + a_2 f_2]$. If $(f_1, f_2) \in \mathcal{F}_\Gamma$, the operator $H^{(\alpha)}[f_1]H^{(\alpha)}[f_2] -$*

---

[9] The word domain is taken to mean, as usual in the context of Hilbert spaces, a linear and dense subset of the space in question.

$H^{(\alpha)}[f_2]H^{(\alpha)}[f_1]$ defined on $\mathcal{D}_{2n}$ is essentially hypermaximally antisymmetric and its unique hypermaximally antisymmetric extension is the operator $iH^{(\alpha)}[(f_1, f_2)]$.

The two corollaries show that the system of operators $H[f]$, $H^{(\alpha)}[f]$ have the same algebraic structure as the set of infinitesimal transformations of the group Γ; recall that the structure of this set was the subject of the theorem in section 5.

By the uniqueness of the hypermaximal extension of an essentially hypermaximal operator, we have that corollary I remains true when the domain $\mathcal{D}_{2n+1}$ is replaced by every larger domain on which the operations in question are still symmetric or antisymmetric. An analogous remark applies to corollary II.

Now we turn to the proof of the theorem stated above. We will obtain this theorem as the consequence of a general theorem about measure-preserving infinitesimal transformations of a C^∞ manifold.

13 A Theorem on Infinitesimal Transformations of a Manifold Into Itself

Let us consider a topological manifold V of l dimensions, enjoying the following properties:

(a) V is C^∞. Thus, every point P of V has at least one open neighborhood v where local coordinates $x_1(P'), \ldots x_l(P')$, $(P' \in v)$, are defined. The equations

$$x_\lambda = x_\lambda(P'), \quad (\lambda = 1, \ldots l)$$

define a homeomorphism of v with a neighborhood in l-dimensional real Euclidean space. Given two coordinate systems defined on the open neighborhoods O', O'' of nonempty intersection

$$x'_\lambda(P'), \ (P' \in O') \quad \text{and} \quad x''_\lambda(P''), \ (P'' \in O''), \quad (P \in O' \cap O'')$$

the coordinate transformation

$$x'_\lambda(P) = \xi_\lambda(x''_1(P), \ldots x''_l(P)), \quad (P \in O' \cap O'') \tag{13.1}$$

is C^∞ and has nonvanishing Jacobian.

(b) There exists on V a countable family of open sets such that every open set of V is a sum of sets in the family (*second axiom of countability*).

(c) *A positive measure of continuous density is defined on V*: it is a completely additive measure μ which is expressed by a continuous and positive density function $\rho(x_1, \ldots x_l)$ on every open set O with a coordinate system $x_\lambda(P)$. For $\mathcal{E} \subset O$, we thus have

$$\mu \mathcal{E} = \int_{\mathcal{E}_x} \rho(x)\, dx_1 \ldots dx_l$$

where $\mathcal{E}_x$ is the image of $\mathcal{E}$ in l-dimensional Euclidean space under the map $P \to x_1(P), \ldots x_l(P)$. The set $\mathcal{E}$ is measurable under μ if $\mathcal{E}_x$ is Lebesgue measurable. Under a coordinate transformation (13.1), ρ transforms as

$$\rho'(x') \left| \frac{\partial(x'_1, \ldots x'_l)}{\partial(x''_1, \ldots x''_l)} \right| = \rho''(x'')$$

Moreover, consider a group $\mathcal{G}$ with real parameter τ ($-\infty < \tau < +\infty$) of bijective transformations of the manifold V into itself, with the following properties:

(d) *The transformations in $\mathcal{G}$ are C^∞ in the points of V and in the parameter τ.* In other words, if g_τ is the transformation of $\mathcal{G}$ which corresponds to the value τ of the parameter, then the map $(\tau, P) \to P' = g_\tau P$ is a continuous map to V from the topological product of V and the real line for all $P \in V$. Further, the local coordinates of P' are C^∞ functions of τ and of the local coordinates of P.

(e) The parameter τ is additive: $g_\tau g_{\tau'} = g_{\tau + \tau'}$.

(f) The transformations in $\mathcal{G}$ preserve the measure μ: for all measurable sets $\mathcal{E}$ of V and all values of τ one has $\mu \mathcal{E} = \mu(g_\tau)$.

Now, consider the Hilbert space $\mathcal{H}$ of complex-valued functions $F(P)$, $G(P)$, ... that are measurable and square integrable on V with respect to the measure μ, with scalar product

$$(F, G) = \int_V \bar{F}(P) G(P)\, d\mu(P).$$

As usual, two functions that only differ in a set of measure zero determine the same element of $\mathcal{H}$.

According to (a), every point $P \in V$ has an open neighborhood $v(P)$ in which local coordinates are defined. Since V satisfies the second axiom of countability, every family of open sets covering V contains a countable subfamily covering V (see for example Lefschetz [17], p. 31, Lindelöf theorem). Thus, there exists a countable sequence of elements $P_1, \ldots P_k, \ldots$ of V such that

$$V = \bigcup_k v(P_k) \qquad (13.2)$$

Denote by $\mathcal{M}_k$ the linearly closed manifold defined in $\mathcal{H}$ by functions which vanish outside $v(P_k)$, and by $\mathcal{N}_k$ the closed manifold spanned by $\mathcal{M}_1, \ldots \mathcal{M}_k$. The $\mathcal{M}_k$, and thus the $\mathcal{N}_k$, are separable. Every element of $\mathcal{H}$ has its orthogonal projection on $\mathcal{N}_k$ as limit when $k \to \infty$. It follows that *the space $\mathcal{H}$ is separable.*

We say that a function $F(P)$ is C^∞ on V if it is a C^∞ function of its coordinates, on every open set where the coordinates are defined. Functions $F(P)$ that are C^∞ on V, and vanish outside compact parts of V, form a subset of $\mathcal{H}$ called $\mathcal{D}$. Returning to the linear manifold $\mathcal{N}_k$ constructed above, one immediately sees that $\mathcal{N}_k \cap \mathcal{D}$ is dense in $\mathcal{D}$. Therefore, *$\mathcal{F}$ is dense in $\mathcal{H}$.*

For $g_\tau \in \mathcal{G}$ and $F(P) \in \mathcal{H}$, we define

$$U_\tau F(P) = F(g_{-\tau} P) . \qquad (13.3)$$

Since the transformations g_τ leave the measure μ invariant, U_τ is a unitary transformation of $\mathcal{H}$. We have the group property $U_\sigma U_\tau = U_{\sigma+\tau}$, and U_0 is the identity transformation. One easily shows that U_τ is strongly continuous in τ: again, the method of section 11 may be used for proving this. The result is that for all values of τ

$$U_\tau = e^{i\tau H} , \qquad (13.4)$$

the operator H being hypermaximally symmetric, with domain D consisting of elements F of $\mathcal{H}$ for which the limit

$$\lim_{\tau \to 0} \tau^{-1}[U_\tau F - F] \qquad (13.5)$$

exists with respect to the norm of $\mathcal{H}$.

Let us show that *$\mathcal{D}$ is contained in the domain D of H*. A function $F \in \mathcal{D}$ is different from zero only on a compact set on V, which can be covered by a finite number of open sets endowed with coordinates. On each such set, F is a C^∞ function of the coordinates, and for τ sufficiently small, g_τ is a C^∞ transformation in these coordinates and in τ. Thus, the limit (13.5) exists, and we have $F \in D$.

Our goal is to establish the following theorem:

THEOREM: *The restriction of the operator H to the domain $\mathcal{D}$ is essentially hypermaximally symmetric.*

The proof will be carried out in several stages. Let us first establish the following lemma.

LEMMA I: *In a Hilbert space $\mathcal{H}$, let A be a hypermaximally symmetric operator of domain D_0 and $\mathcal{D}_0$ a linear subset of D_0, dense in $\mathcal{H}$. In order that the restriction of A to $\mathcal{D}_0$ be essentially hypermaximally symmetric, it is necessary and sufficient that, for a complex number ζ with nonzero real part, the operator $A + i\zeta$ transforms $\mathcal{D}_0$ into a dense set of $\mathcal{H}$. If this condition is satisfied for one number ζ with nonzero real part, it is satisfied for all such numbers.*

The last part is shown by considering the spectral representation of A, in obvious notation,

$$A = \int_{-\infty}^{+\infty} \lambda \, dE(\lambda) \, .$$

For ζ_1, ζ_2 complex with nonzero real part, the operator

$$\frac{A + i\zeta_1}{A + i\zeta_2} = \int_{-\infty}^{+\infty} \frac{\lambda + i\zeta_1}{\lambda + i\zeta_2} \, dE(\lambda)$$

is defined on all of $\mathcal{H}$, is bounded and has inverse $(A + i\zeta_2)(A + i\zeta_1)^{-1}$ which enjoys the same properties. It follows that if the set $(A + i\zeta_2)\mathcal{D}_0$ is dense in $\mathcal{H}$, the same is true for

$$(A + i\zeta_1)\mathcal{D}_0 = \frac{A + i\zeta_1}{A + i\zeta_2}(A + i\zeta_2)\mathcal{D}_0 \, .$$

The first part of the lemma can also be shown using the spectral representation of A. But it arises immediately from the last part, combined with the theory of symmetric transformation of Hilbert space. Indeed, by this theory, the restriction of A to $\mathcal{D}_0$ admits either one or an infinity of single hypermaximal extension(s), depending on whether the sets $(A \pm i)\mathcal{D}_0$ are dense in $\mathcal{H}_0$ or not (see Stone[30] p. 339, theorem 9.3 and the remarks that follow).

We now consider the topological manifold $\tilde{V}$, topological product of the manifold V with the real line $-\infty < \sigma < +\infty$. In $\tilde{V}$, we define the measure ν, product of the measure μ of V and the ordinary measure on the real line. If $\mathcal{E}$ is a measurable set of V and e a measurable set of the real line, the product set $\mathcal{E} \times e$ is measurable under ν and

$$\nu(\mathcal{E} \times e) = \mu(\mathcal{E}) \cdot m(e) \, ,$$

$m(e)$ being the Lebesgue measure of e. Like V, $\tilde{V}$ is a C^∞ manifold satisfying the second axiom of countability, and the measure ν defined on $\tilde{V}$ admits a continuous density; thus we have properties analogous to *(a), (b), (c)*. We denote by $\tilde{\mathcal{D}}$ the set of C^∞ functions on $\tilde{V}$, vanishing outside compact sets. Denoting the points of $\tilde{V}$ by (σ, P), (σ real, $P \in V$), we define the group $\tilde{\mathcal{G}}$ of transformations $\tilde{g}_\tau$ satisfying[t32]

$$\tilde{g}_\tau(\sigma, P) = (\sigma + \tau, P)$$

These transformations are C^∞ in the points of $\tilde{V}$ and in the parameter τ; they form a group and leave the measure μ invariant; thus, they satisfy properties analogous to (d), (e), (f).

Let us consider the Hilbert space $\tilde{\mathcal{H}}$ of complex-valued functions $\tilde{F}(\sigma, P)$, measurable and square integrable on $\tilde{V}$ by the measure ν. In this space, the group $\tilde{\mathcal{G}}$ defines the strongly continuous group of unitary transformations

$$\tilde{U}_\tau \tilde{F}(\sigma, P) = \tilde{F}(\sigma - \tau, P) \ .$$

We again have

$$\tilde{U} = e^{i\tau \tilde{H}} \ ,$$

$\tilde{H}$ being a hypermaximally symmetric operator with domain containing $\tilde{\mathcal{D}}$; on $\tilde{\mathcal{D}}$ (in fact, on a larger domain), we have

$$\tilde{H} = i \frac{\partial \tilde{F}}{\partial \sigma}$$

Now we establish a second lemma:

LEMMA II: *The restriction of the operator $\tilde{H}$ to the domain $\tilde{\mathcal{D}}$ is essentially hypermaximally symmetric.*

According to lemma I, it is enough to show that the set $(\tilde{H}+i)\tilde{\mathcal{D}}$ is dense in $\tilde{\mathcal{H}}$. Assume the contrary, that there is an element $\tilde{F}(\sigma, P)$ of $\tilde{\mathcal{H}}$ orthogonal to all elements of $(\tilde{H}+i)\tilde{\mathcal{D}}$. It satisfies

$$\int_{\tilde{V}} \bar{\tilde{F}}(\sigma, P) \left(\frac{\partial}{\partial \sigma} + 1 \right) \tilde{G}(\sigma, P) \, d\sigma \, d\mu(P) = 0 \ ,$$

for all $\tilde{G} \in \tilde{\mathcal{D}}$. Since the map $\tilde{G}(\sigma, P) \to e^{-\sigma} \tilde{G}(\sigma, P)$ transforms $\tilde{\mathcal{D}}$ into itself, we also have for all $\tilde{G} \in \tilde{D}$

$$\int_{\tilde{V}} \bar{\tilde{F}}(\sigma, P) \left(\frac{\partial}{\partial \sigma} + 1 \right) e^{-\sigma} \tilde{G}(\sigma, P) \, d\sigma \, d\mu(P) =$$
$$\int_{\tilde{V}} \bar{\tilde{F}} e^{-\sigma} \frac{\partial}{\partial \sigma} \tilde{G} \, d\sigma \, d\mu(P) = 0 \ .$$

Take $\tilde{G}(\sigma, P) = \varphi(\sigma) G(P)$ where $G(P) \in \mathcal{D}$ and $\varphi(\sigma)$ is a C^∞ function of σ, zero outside a compact set. The Fubini theorem gives

$$\int_{-\infty}^{+\infty} e^{-\sigma} \frac{d\varphi}{d\sigma} \, d\sigma \int_V \bar{\tilde{F}}(\sigma, P) \tilde{G}(P) \, d\mu(P) = 0 \ . \qquad (13.6)$$

Applying a procedure well known in the Calculus of Variations, we let $\varphi(\sigma)$ uniformly tend to the function

$$\varphi_0(\sigma) = \begin{cases} 0 & \text{for } -\infty < \sigma \leq a_1 \\ (a_2 - a_1)^{-1}(\sigma - a_1) & \text{for } a_1 \leq \sigma \leq a_2 \\ 1 & \text{for } a_2 \leq \sigma \leq b_1 \\ (b_2 - b_1)^{-1}(b_2 - \sigma) & \text{for } b_1 \leq \sigma \leq b_2 \\ 0 & \text{for } b_2 \leq \sigma < +\infty, \end{cases}$$

such that $d\varphi/d\sigma$ uniformly converges to $d\varphi_0/d\sigma$. One can take the limit in the integral (13.6) and one finds

$$\frac{1}{a_2 - a_1} \int_{a_1}^{a_2} e^{-\sigma} d\sigma \int_V \bar{\tilde{F}} G \, d\mu(P) = \frac{1}{b_2 - b_1} \int_{b_1}^{b_2} e^{-\sigma} d\sigma \int_V \bar{\tilde{F}} G \, d\mu(P).$$

Thus, since c does not depend on a_1, a_2,

$$\int_{a_1}^{a_2} e^{-\sigma} d\sigma \int_V \bar{\tilde{F}}(\sigma, P) \tilde{G}(P) \, d\mu(P) = c(a_2 - a_1)$$

and by differentiation, for almost all values of σ

$$\int_V \bar{\tilde{F}}(\sigma, P) \tilde{G}(P) \, d\mu(P) = c e^{\sigma}. \tag{13.7}$$

But the integral

$$\int_{-\infty}^{+\infty} \psi(\sigma) \, d\sigma \int_V \bar{\tilde{F}}(\sigma, P) \tilde{G}(P) \, d\mu(P) \tag{13.8}$$

must be convergent for all measurable and square integrable functions $\psi(\sigma)$. Therefore, c must vanish in (13.7). The integral (13.8) is then zero; since we may arbitrarily choose $G(P) \in \mathcal{D}$ and $\psi(\sigma)$ measurable, $\tilde{F}$ must be the zero element of $\tilde{\mathcal{H}}$. We have shown that the set $(\tilde{H} + i)\tilde{\mathcal{D}}$ is dense in $\tilde{\mathcal{H}}$. Lemma II is thus established.

We return to the group $\mathcal{G}$ of transformations g_τ of the manifold V into itself. We now use this group for defining the following transformation T of $\tilde{V}$ into itself

$$T(\sigma, P) = (\sigma, g_\sigma P).$$

It is a C^∞ homeomorphism, leaving the measure ν invariant. In $\tilde{\mathcal{H}}$, T gives rise to the unitary transformation

$$W\tilde{F}(\sigma, P) = \tilde{F}(\sigma, g_{-\sigma} P)$$

which maps the set $\tilde{\mathcal{D}}$ into itself.

The group $\tilde{\mathcal{G}}$ transformed by T is called $\tilde{\mathcal{G}}'$, and is given by transformations

$$\tilde{g}'_\tau(\sigma, P) = T\tilde{g}_\tau T^{-1}(\sigma, P) = (\sigma + \tau, g_\tau P) \,.$$

To the group $\tilde{\mathcal{G}}'$ corresponds a group of unitary transformation in $\tilde{\mathcal{H}}$

$$\tilde{U}'_\tau = W\tilde{U}_\tau W^{-1} = e^{i\tau \tilde{H}'} \,, \quad \tilde{U}'_\tau \tilde{F}(\sigma, P) = \tilde{F}(\sigma - \tau, g_{-\tau} P) \,,$$

where the hypermaximally symmetric operator $\tilde{H}' = W\tilde{H}W^{-1}$ has as domain the transform by W of the domain of $\tilde{H}$; thus, it is defined on $\tilde{\mathcal{D}}$, in particular. One finds for the elements of $\tilde{\mathcal{D}}$

$$\tilde{H}'\tilde{F}(\sigma, P) = i\frac{\partial \tilde{F}}{\partial \sigma} + H_P \tilde{F}(\sigma, P) \,, \tag{13.9}$$

where H_P denotes the operator H defined in (13.3), (13.4) applied to $\tilde{F}(\sigma, P)$, which is now considered as a function of P for every σ, that is, as an element of $\mathcal{D}$. Lemma III, which follows, is a consequence of lemma II.

LEMMA III: *The restriction of the operator $\tilde{H}'$ to the domain $\tilde{\mathcal{D}}$ is essentially hypermaximally symmetric.*

We now pass to the final stage of our proof. Let J be the unitary transformation of $\tilde{\mathcal{H}}$ defined by the Fourier integral

$$J\tilde{F}(\sigma, P) = \frac{1}{\sqrt{2\pi}} \int_{-\infty}^{+\infty} e^{i\sigma\sigma'} \tilde{F}(\sigma', P) d\sigma' \,.$$

From (13.9) one finds the relation

$$J\tilde{H}'J^{-1}\tilde{F}(\sigma, P) = (\sigma + H_P)\tilde{F}(\sigma, P) \tag{13.10}$$

which is valid for elements with domain $J\tilde{\mathcal{D}}$; H_P has the same meaning as in (13.9). According to lemma III, $(\tilde{H}' + i)\tilde{\mathcal{D}}$ is dense in $\tilde{\mathcal{H}}$. Thus, the domain $J(H' + i)\tilde{\mathcal{D}}$ is dense in $\mathcal{H}$. We conclude that the domain $(H + i)\mathcal{D}$ is dense in $\mathcal{H}$.

Let $\mathcal{M}$ be the linearly closed manifold spanned by the domain $(H + i)\mathcal{D}$, and let us suppose that $\mathcal{H}$ contains a nonzero element $F(P)$ orthogonal to $\mathcal{M}$. For λ real, consider the linear bounded operator

$$\frac{H + \lambda + i}{H + i} = I + A_\lambda$$

where I is the identity transformation in $\mathcal{H}$, and A_λ is a bounded operator of bound a_λ. Using the spectral representation of H, one immediately sees that for $|\lambda| \leq \lambda_0(\eta)$ sufficiently small, a_λ remains smaller than an arbitrary positive number η, which we take smaller than $1/2$. Now, for the element $F(P)$ that we assumed exists, and for all $G \in \mathcal{M}$, we have

$$\begin{aligned} |(F, (I + A_\lambda))| &= |(F, A_\lambda G)| \\ &\le a_\lambda ||F|| \cdot ||G|| \le \eta ||F|| \cdot ||G|| \\ &\le \frac{\eta}{1-\eta} ||F|| \cdot ||(I + A_\lambda)G|| \end{aligned} \qquad (13.11)$$

provided $|\lambda| \le \lambda_0(\eta)$. In $\tilde{\mathcal{H}}$, we define the element

$$\tilde{F}(\sigma, P) = \begin{cases} F(P) & \text{for} \quad |\sigma| \le \lambda_0(\eta), \\ 0 & \text{for} \quad |\sigma| > \lambda_0(\eta). \end{cases} \qquad (13.12)$$

If $\tilde{G} \in \tilde{\mathcal{D}}$, set $\tilde{G}' = J\tilde{G}$; then using (13.10), we can calculate the scalar product in $\mathcal{H}$:

$$(\tilde{F}, J(\tilde{H}' + i)\tilde{G}) = \int_{-\lambda_0(\eta)}^{\lambda_0(\eta)} d\sigma \int_V \bar{F}(P) \left\{ (H_P + \sigma + i)\tilde{G}'(\sigma, P) \right\} d\mu(P).$$

For every value of σ, $\tilde{G}'(\sigma, P)$ defines an element G'_σ of $\mathcal{D}$. The preceding equality is still

$$(\tilde{F}, J(\tilde{H}' + i)\tilde{G}) = \int_{-\lambda_0(\eta)}^{\lambda_0(\eta)} (F, (H + \sigma + i)G'_\sigma) \, d\sigma.$$

The scalar product which appears in the second integral is taken in the space $\mathcal{H}$. If $G''_\sigma = (H + i)G'_\sigma \in \mathcal{M}$, we find using (13.12) and the Schwarz inequality

$$\begin{aligned} |(\tilde{F}, J(\tilde{H}' + i)\tilde{G})| &= \int_{-\lambda_0(\eta)}^{\lambda_0(\eta)} (F, (I + A_\sigma)G''_\sigma) d\sigma \\ &\le \frac{\eta}{1-\eta} ||F|| \int_{-\lambda_0(\eta)}^{\lambda_0(\eta)} ||(H + \sigma + i)G'_\sigma|| \, d\sigma \\ &\le \frac{\eta}{1-\eta} \sqrt{2\lambda_0(\eta)} ||F|| \cdot \left\{ \int_{-\lambda_0(\eta)}^{\lambda_0(\eta)} ||(H + \sigma + i)G'_\sigma||^2 \, d\sigma \right\}^{1/2} \end{aligned}$$

The last integral is nothing else than the square of the norm of $J(\tilde{H}' + i)\tilde{G}$ in $\tilde{\mathcal{H}}$, and $\sqrt{2\lambda_0(\eta)} ||F||$ is the norm of $\tilde{F}$ in $\tilde{\mathcal{H}}$. We have thus shown that, in $\tilde{\mathcal{H}}$, the element (13.12) satisfies

$$(\tilde{F}, \tilde{G}) \le \frac{\eta}{1-\eta} ||\tilde{F}|| \cdot ||\tilde{G}||$$

for all elements $\tilde{G}$ of the domain $J(\tilde{H}' + i)\tilde{\mathcal{D}}$. This cannot be so since the domain is dense in $\tilde{\mathcal{H}}$. Consequently, the manifold $\mathcal{M}$ is none other than the entire space $\mathcal{H}$. Therefore, the domain $(H+i)\mathcal{D}$ is dense in $\mathcal{H}$, and by lemma I the restriction of H to $\mathcal{D}$ is essentially hypermaximal. The theorem is then established.

14 Application to Operators $H[f]$, $H^{(\alpha)}[f]$

Let us now return to the operators $H[f]$, $H^{(\alpha)}[f]$ for $f \in \mathcal{F}_\Gamma$. Using the theorem of the previous section, we should show that the restrictions of these operators to domains $\mathcal{D}_{2n+1}$ and $\mathcal{D}_{2n}$, respectively, are essentially hypermaximally symmetric.

For $f \in \mathcal{F}_\Gamma$, denote by $\gamma_\tau^{(f)}$ the transformations of the one-parameter subgroup Γ generated by the infinitesimal transformation $X[f]$. We apply the theorem of the previous section, taking the space (s, p, q) as our C^∞ manifold V, with measure defined by the volume element $ds\, dp_1, \ldots dp_n\, dq_1 \ldots dq_n$ and the transformations $\gamma_\tau^{(f)}$ as our one-parameter group. For the operator H we have $H[f]$, for the domain $\mathcal{D}$ we have $\mathcal{D}_{2n+1}$. The restriction of $H[f]$ to $\mathcal{D}_{2n+1}$ is thus essentially hypermaximal.

Further, take for V the space (p, q) with measure defined by $dp_1, \ldots dp_n\, dq_1 \ldots dq_n$, and for group action use $\gamma_\tau^{(f)}$ on the variables p_j, q_j. For the operator H we now have $H^{(0)}[f]$, for $\mathcal{D}$ we have $\mathcal{D}_{2n}$. Thus, it is for the restriction of $H^{(0)}[f]$ to $\mathcal{D}_{2n}$ that we find the essentially hypermaximal property.

Finally, let α be real and nonzero. We now take for V the manifold obtained from the space (s, p, q) by identifying points s, p_j, q_j and s', p'_j, q'_j for which

$$s' \equiv s \pmod{2\pi/\alpha}, \; p'_j = p_j, \; q'_j = q_j, \; (j = 1, \ldots n). \qquad (14.1)$$

This manifold is the topological product of the space (p, q) and a circle.[33] A transformation γ of the group Γ leaves the relations (14.1) invariant. Thus, it defines a transformation of V into itself, which we also denote by γ; this transformation is C^∞. The Hilbert space of functions that are measurable and square integrable on V is again called $\mathcal{H}$. Denote the points of V by P. Let $F(P) \in \mathcal{H}$, $\gamma \in \Gamma$; we define as usual

$$\mathcal{U}_\gamma F(P) = F(\gamma^{-1} P).$$

Now the domain $\mathcal{D}$ is formed by C^∞ functions $\phi(s, p_1, \ldots p_n, q_1 \ldots q_n)$, periodic with period $2\pi/\alpha$, and which all vanish when the point p_j, q_j is outside a compact set of the space (p, q).

Let us consider the one-parameter group of transformations $\gamma_\tau^{(f)}$ on V and let $\mathcal{U}_{\gamma_\tau^{(f)}} = e^{i\tau K}$. The hypermaximally symmetric operator K has, for

$\phi \in \mathcal{D}$, the same form as (11.2)

$$K\phi = i\left(f - \sum p_j \frac{\partial f}{\partial p_j}\right)\frac{\partial \phi}{\partial s} + i\sum \left(\frac{\partial f}{\partial q_j}\frac{\partial \phi}{\partial p_j} - \frac{\partial f}{\partial p_j}\frac{\partial \phi}{\partial q_j}\right). \quad (14.2)$$

By the theorem of the previous paragraph, the restriction of K to $\mathcal{D}$ is essentially hypermaximal.

Let $\varphi(\omega) = \varphi(p_1, \ldots p_n, q_1, \ldots q_n)$ be an element of the Hilbert space $\mathcal{H}_{2n}$ defined in section 8. The map

$$\varphi(\omega) \to \sqrt{\frac{\alpha}{2\pi}} e^{-i\alpha ks} \varphi(\omega), \quad (k = 0, \pm 1, \pm 2, \ldots) \quad (14.3)$$

is a bijective map A_k, conserving the scalar product, from $\mathcal{H}_{2n}$ to a linearly closed manifold $\mathcal{M}_k$ in $\mathcal{H}$. The manifolds $\mathcal{M}_k$ are invariant under unitary transformations $\mathcal{U}_\gamma$, $\gamma \in \Gamma$. They are pairwise orthogonal and span the entire space $\mathcal{H}$, which is thus their direct sum. If $\phi(s, \omega) \in \mathcal{H}$, we have by expansion in Fourier series

$$\phi(s, \omega) = \sum_{-\infty}^{+\infty} A_k \sqrt{\frac{\alpha}{2\pi}} \int_0^{2\pi/a} e^{i\alpha k s'} \phi(s', \omega) \, ds' \, .$$

We find that the orthogonal projection of the domain $\mathcal{D}$ in $\mathcal{H}$ onto $\mathcal{M}_k$ is none other than $A_k \mathcal{D}_{2n}$. Furthermore, by (14.2), the operator K maps every domain $A_k \mathcal{D}_{2n}$ into itself. We know by the previous section that $(K+i)\mathcal{D}$ is dense in $\mathcal{H}$. Thus, for every value of k, $(K+i)A_k\mathcal{D}_{2n}$ is dense in $\mathcal{M}_k$, and the restriction of K to $A_k\mathcal{D}_{2n}$ is essentially hypermaximally symmetric there. But (12.1), (14.2), (14.3) immediately show that for $\varphi \in \mathcal{D}_{2n}$

$$A_k^{-1} K A_k \varphi = H^{(k\alpha)}[f]\varphi \, .$$

We see that the restrictions of the operators $H^{(k\alpha)}[f]$, $(k = 0, \pm 1, \ldots)$ to the domain $\mathcal{D}_{2n}$ are essentially hypermaximally symmetric. The theorem of section 12 is thus fully established.

15 A Lie Algebra of Operators

Let us now restrict attention to elements f of the algebra $\mathcal{F}$ defined in section 6. We have seen that the set $X[\mathcal{F}]$ of the corresponding infinitesimal transformations forms a Lie algebra isomorphic to $\mathcal{F}$. The families of hypermaximally antisymmetric operators $-iH[f]$, $-iH^{(\alpha)}[f]$, $(f \in \mathcal{F})$, which represent these infinitesimal transformations in $\mathcal{R}$, $\mathcal{R}^{(\alpha)}$, must have analogous algebraic structure. However, there is a new difficulty; the sums and commutators of the operators we consider only make sense on sufficiently restricted domains of definition.

This difficulty concerns the definition of a Lie algebra of hypermaximally antisymmetric operators. A satisfactory definition has recently been proposed by Segal (Segal[28]) [h] It can be stated as follows:

A system $\mathcal{A}$ of hypermaximally antisymmetric operators in the Hilbert space $\mathcal{H}$ constitutes a Lie algebra of operators if it has the following properties:

(a) There exists a domain $\mathcal{D}$, dense in $\mathcal{H}$, contained in the domain of definition of all $A \in \mathcal{A}$, invariant for all $A \in \mathcal{A}$ (that is, $A\mathcal{D} \subset \mathcal{D}$ for all $A \in \mathcal{A}$) and such that the restriction of all $A \in \mathcal{A}$ to the domain $\mathcal{D}$ is essentially hypermaximally antisymmetric.

(b). Let A', A'' be any two elements of $\mathcal{A}$, and a', a'' two real numbers; there exist elements A_1, A_2 in $\mathcal{A}$ such that the equalities

$$(a'A' + a''A'')\varphi = A_1\varphi$$
$$(A'A'' - A''A')\varphi = A_2\varphi$$

are satisfied for all $\varphi \in \mathcal{D}$.

Note that with this definition, linear combinations with real coefficients, and the commutator of any two elements of $\mathcal{A}$

- are defined on $\mathcal{D}$
- are essentially hypermaximally antisymmetric there, and
- both have as hypermaximal extension a uniquely determined element of $\mathcal{A}$. [t34]

To illustrate the difficulties that can arise when one attempts to prove that a system of operators constitutes a Lie algebra, let us mention, as did Segal, that according to von Neumann, one can find two essentially hypermaximally symmetric operators which have a common domain where their sum is *not* essentially hypermaximal. [t35]

Having said this, the set $\mathcal{F}$ is closed under linear combination with real coefficients and under Poisson brackets, being a Lie algebra. By the corollaries of section 12, we then immediately have

THEOREM: *The set $\mathcal{R}[\mathcal{F}]$ of hypermaximally antisymmetric $-iH[f]$, ($f \in \mathcal{F}$), constitutes a Lie algebra of operators. For all real α, the same holds for the set $\mathcal{R}^{(\alpha)}[\mathcal{F}]$ of hypermaximally antisymmetric $-iH^{(\alpha)}[f]$, ($f \in \mathcal{F}$). The*

---

[h]With the kindness of M. SEGAL, we were able to consult this work before publication.

correspondences $f \to -iH[f]$, $f \to -iH^{(\alpha)}[f]$ *are isomorphisms of* $\mathcal{F}$ *with the Lie algebras* $\mathcal{R}[\mathcal{F}]$, $\mathcal{R}^{(\alpha)}[\mathcal{F}]$.

Similarly, the correspondences $X[f] \to -iH[f]$, $X[f] \to -iH^{(\alpha)}[f]$ between infinitesimal transformations in Γ and hypermaximally antisymmetric operators which represent them in $\mathcal{R}$, $\mathcal{R}^{(\alpha)}$ are isomorphisms of the Lie algebra $X[F]$ with the Lie algebras $R[\mathcal{F}]$, $\mathcal{R}^{(\alpha)}[\mathcal{F}]$.

These properties clearly remain true if one replaces $\mathcal{F}$ by any other Lie algebra contained within $\mathcal{F}_\Gamma$, that is, by any other subset of $\mathcal{F}_\Gamma$ which is closed under linear combination with real coefficients and under Poisson brackets.

CHAPTER V

Irreducibility of the Representations $\mathcal{R}^{(\alpha)}$.

The hypermaximally antisymmetric operators in the representation $\mathcal{R}^{(\alpha)}$ ($\alpha \neq 0$) which correspond to infinitesimal transformations in Γ are cast in new form. This form serves two purposes: reducing $\mathcal{R}^{(\alpha)}$ ($\alpha \neq 0$) to a representation of the subgroup L of Γ, and proving irreducibility of the representation $\mathcal{R}^{(\alpha)}$ of the group Γ. Further, we show that $\mathcal{R}^{(0)}$ is an irreducible representation of L and *a fortiori* of Γ.

16 New Expression for the Operators $H^{(\alpha)}[f]$ for $\alpha \neq 0$

As we will discuss in greater detail in the following chapter, the operators $\frac{1}{\alpha} H^{(\alpha)}[f]$ present a remarkable analogy with the operators of Quantum Mechanics, particularly with their commutation relations. However, they differ by the fact that the Hilbert space $\mathcal{H}_{2n}$ is not irreducible for all one-parameter groups generated by the $2n$ operators $\frac{1}{\alpha} H^{(\alpha)}[p_j]$, $\frac{1}{\alpha} H^{(\alpha)}[q_j]$ ($j = 1, \ldots n$). We mention again that in $\mathcal{H}_{2n}$, the identity transformation is represented by distinct operators, which commute with these groups. For the proof of irreducibility that we have in mind, it is useful to construct such operators; they are constructed by a unitary transformation W of $\mathcal{H}_{2n}$ such that the transforms of $\frac{1}{\alpha} H^{(\alpha)}[p_j]$, $\frac{1}{\alpha} H^{(\alpha)}[q_j]$ by W act on only n of the $2n$ independent variables $p_1, \ldots p_n, q_1 \ldots q_n$. These n variables can be arbitrarily chosen among the p_j, q_j; as we shall see, if one adopts the following transformation W, they are actually the q_j: for functions $\varphi(p_1, \ldots p_n, q_1, \ldots q_n)$, measurable and square integrable on the space (p, q) we define the transformation

$$W\varphi(p_1, \ldots p_n, q_1 \ldots q_n) = \qquad (16.1)$$
$$\left(\frac{\alpha}{2\pi}\right)^{n/2} \int_{-\infty}^{+\infty} dv_1 \ldots \int_{-\infty}^{+\infty} dv_n \, e^{i\alpha \sum_1^n v_j p_j} \varphi(v_1, \ldots v_n, q_1 - v_1, \ldots q_n - v_n) \, .$$

By the Plancherel theorem, we immediately see that the transformation is unitary; its defining Fourier integral converges in the norm of $\mathcal{H}_{2n}$. On the other hand, consider the following hypermaximally symmetric operators of a well-known type, for nonzero α (see for example Stone [30] p. 441, theorem 10.9):

$$A_j \varphi(p, q) = \frac{1}{\alpha i} \frac{\partial}{\partial q_j} \varphi(p, q) \, ; \quad B_j \varphi(p, q) = q_j \varphi(p, q) \qquad (16.2)$$

44

$$A'_j \varphi(p,q) = \frac{1}{\alpha i} \frac{\partial}{\partial p_j} \varphi(p,q) \; ; \quad B'_j \varphi(p,q) = p_j \varphi(p,q) \; . \quad (j = 1, \ldots n) \, . \tag{16.3}$$

Their domains of definition all contain the set Δ_{2n} of C^∞ functions on the space (p,q) that are bounded and stay bounded after every finite succession of partial differentiations and multiplication by $p_1, \ldots p_n, q_1, \ldots q_n$. As usual, $\mathcal{D}_{2n}$ denotes the set of C^∞ functions on the space (p,q), vanishing outside compact sets; we have $\mathcal{D}_{2n} \subset \Delta_{2n} \subset \mathcal{H}_{2n}$. The domain Δ_{2n} is invariant under the operators A_j, B_j, A'_j, B'_j as well as under the transformation W. For the elements in this domain, one easily checks the following relations, by partial integration and differentiation under the integral sign,

$$P_j = W A_j W^{-1} = A_j \; ; \quad Q_j = W B_j W^{-1} = B_j - A'_j \; ;$$
$$(j = 1, \ldots n) \tag{16.4}$$
$$P'_j = W A'_j W^{-1} = A_j - B'_j \; ; \quad Q'_j = W B'_j W^{-1} = A'_j \, .$$

The operators P_j, Q_j, P'_j, Q'_j that are defined here have domains of definition that all contain Δ_{2n}, thus $\mathcal{D}_{2n}$. *Their restrictions to $\mathcal{D}_{2n}$ are essentially hypermaximal*. To convince oneself that this is true, it suffices to note that by (12.1), for the elements of $\mathcal{D}_{2n}$ and the relevant value of α,

$$P_j = \frac{1}{\alpha} H^{(\alpha)}[p_j] \, . \quad Q_j = \frac{1}{\alpha} H^{(\alpha)}[q_j] \, . \tag{16.5}$$

According to the theorem of section 12, the restrictions of P_j, Q_j to $\mathcal{D}_{2n}$ are therefore essentially hypermaximal; by symmetry in the variables p_j, q_j, the same conclusion holds for P'_j, Q'_j. Henceforth, we call P_j, Q_j, P'_j, Q'_j the hypermaximal extensions defined in (16.4).

Let us find the commutators of the operators we have just introduced. For the elements of $\mathcal{D}_{2n}$, we have

$$Q_j P_k - P_k Q_j = \frac{1}{\alpha^2} \left\{ H^{(\alpha)}[q_j] H^{(\alpha)}[p_k] - H^{(\alpha)}[p_k] H^{(\alpha)}[q_j] \right\}$$
$$= \frac{i}{\alpha^2} H^{(\alpha)}[(q_j, p_k)] = \frac{i}{\alpha^2} H^{(\alpha)}[\delta_{jk}] = \frac{i}{\alpha} \delta_{jk}$$

where δ_{jk} is the Kronecker delta symbol; similarly,

$$Q_j Q_k - Q_k Q_j = 0 \, , \; P_j P_k - P_k P_j = 0 \, .$$

By symmetry

$$Q'_j P'_k - P'_k Q'_j = \frac{i}{\alpha} \delta_{jk} \, , \; Q'_j Q'_k - Q'_k Q'_j = 0 \, , \; P'_j P'_k - P'_k P'_j = 0 \, .$$

Moreover, these formulas follow immediately from the use of relations (16.2), (16.3) and (16.4) which also give

$$Q'_j Q'_k - Q'_k Q'_j = 0 , \quad P_j Q'_k - Q'_k P_j = 0 ,$$
$$Q_j P'_k - P'_k Q_j = 0 , \quad P_j P'_k - P'_k P_j = 0 .$$

For all real α, we have introduced two systems of $2n$ operators, the system $\mathcal{S}$ of operators $P_1, \ldots P_n, Q_1, \ldots Q_n$, and the system $\mathcal{S}'$ of $P'_1, \ldots P'_n, Q'_1, \ldots Q'_n$. As the operators of $\mathcal{S}$ commute with those of $\mathcal{S}'$, commutation relations are identical with the well-known commutation rules of Quantum Mechanics both in $\mathcal{S}$ and $\mathcal{S}'$ (provided one identifies $2\pi/\alpha$ with Planck's constant). We shall return to this point in the following chapter.

Now we express the operator $H^{(\alpha)}[f]$, for all $f \in \mathcal{F}_\Gamma$ and all $\alpha \neq 0$, in terms of functions of operators in the systems $\mathcal{S}$ and $\mathcal{S}'$ for a single value of α. By (16.4), we have on the domain $\mathcal{D}_{2n}$

$$\begin{aligned} A_j &= P_j & A'_j &= Q'_j , \\ B_j &= Q_j + Q'_j , & B_j &= P_j - P'_j . \end{aligned} \qquad (16.6)$$

On the other hand, by (12.1) for α real, $f \in \mathcal{F}_\Gamma$, $\varphi \in \mathcal{D}_{2n}$, we have

$$\frac{1}{\alpha} H^{(\alpha)}[f]\varphi = \left(f - \sum P_j \frac{\partial f}{\partial p_j}\right)\varphi - \frac{1}{i\alpha} \sum \left(\frac{\partial f}{\partial q_j}\frac{\partial \varphi}{\partial p_j} - \frac{\partial f}{\partial p_j}\frac{\partial \varphi}{\partial q_j}\right) .$$

Assume that

$$\frac{\partial f}{\partial p_j} = f_{p_j}(p_1, \ldots p_n, q_1 \ldots q_n) , \quad \frac{\partial f}{\partial q_j} = f_{q_j}(p_1, \ldots p_n, q_1 \ldots q_n) .$$

and let us introduce (16.2), (16.3) and (16.6). We find that

$$\begin{aligned} \frac{1}{\alpha} H^{(\alpha)}[f]\varphi = \{ &f(P_1 - P'_1, \ldots P_n - P'_n, Q_1 + Q'_1, \ldots Q_n + Q'_n) \\ &+ f(P_1 - P'_1, \ldots P_n - P'_n, Q_1 + Q'_1, \ldots Q_n + Q'_n)P'_j \qquad (16.7) \\ &- f(P_1 - P'_1, \ldots P_n - P'_n, Q_1 + Q'_1, \ldots Q_n + Q'_n)Q'_k \} \varphi . \end{aligned}$$

Given that

$$\frac{\partial f}{\partial q_j}\frac{\partial \varphi}{\partial p_j} - \frac{\partial f}{\partial p_j}\frac{\partial \varphi}{\partial q_j} = \frac{\partial}{\partial p_j}\left(\frac{\partial f}{\partial q_j}\varphi\right) - \frac{\partial}{\partial q_j}\left(\frac{\partial f}{\partial p_j}\varphi\right)$$

one also has

$$\begin{aligned} \frac{1}{\alpha} H^{(\alpha)}[f]\varphi = \{ &f(P_1 - P'_1, \ldots P_n - P'_n, Q_1 + Q'_1, \ldots Q_n + Q'_n) \\ &+ \sum_j P'_j f_{p_j}(P_1 - P'_1, \ldots P_n - P'_n, Q_1 + Q'_1, \ldots Q_n + Q'_n) \quad (16.8) \end{aligned}$$

$$-\sum_j Q'_j f_{q_j}(P_1 - P'_1, \ldots P_n - P'_n, Q_1 + Q'_1, \ldots Q_n + Q'_n)\}\varphi.$$

One can also take half the sum of expressions (16.7) and (16.8); every term in the expression so obtained is individually a symmetric operator on $\mathcal{D}_{2n}$. Note that, according to the commutation relations below, the operators $P_1 - P'_1, \ldots, P_n - P'_n, Q_1 + Q'_1, \ldots Q_n + Q'_n$ commute pairwise. Functions of operators appearing in (16.7) and (16.8) are thus perfectly well-defined (see von Neumann [21]). We have obtained the following result:

THEOREM: *For $f \in \mathcal{F}_\Gamma$ and $\alpha \neq 0$ one has the equality*

$$\frac{1}{\alpha} H^{(\alpha)}[f] = f(P - P', Q + Q') + \sum_j$$
$$\{P'_j f_{p_j}(P - P', Q + Q') - Q'_j f_{q_j}(P - P', Q + Q')\} \quad (16.9)$$
$$= f(P - P', Q + Q') + \sum_j$$
$$\{f_{p_j}(P - P', Q + Q')P'_j - f_{q_j}(P - P', Q + Q')Q'_j\}$$

where $H^{(\alpha)}[f]$ and the operators P_j, Q_j, P'_j, Q'_j are taken for the same real value of α. This equality is valid on the domain $\mathcal{D}_{2n}$ on which the operators it relates are essentially hypermaximally symmetric. Thus, it applies to hypermaximal extensions of these operators as well.

17 Reduction of the Representation $\mathcal{R}^{(\alpha)}$ of the Subgroup L for $\alpha \neq 0$

In order to establish irreducibility of the representation $\mathcal{R}^{(\alpha)}(\alpha \neq 0)$ of the group Γ, it is interesting to represent the subgroup L mentioned in section 4 by a reduction of $\mathcal{R}^{(\alpha)}$.[t36] We will do so now, by introducing a basis of Hermite polynomials in the Hilbert space $\mathcal{H}_{2n}$. We introduce the Hermite functions $h_m(x)$ of one real variable x ($-\infty < x < +\infty$),

$$h_m(x) = e^{x^2/2} \frac{d^m}{dx^m} e^{-x^2}, \quad (m = 0, 1, 2, \ldots).$$

The following properties are well-known

$$xh_m(x) = mh_{m-1}(x) + \frac{1}{2}h_{m+1}(x),$$
$$\frac{d}{dx}h_m(x) = mh_{m-1}(x) - \frac{1}{2}h_{m+1}(x), \quad (17.1)$$
$$\frac{1}{2}\left(x^2 - \frac{d^2}{dx^2}\right)h_m(x) = \left(m + \frac{1}{2}\right)h_m(x),$$

as well as the fact that the $h_m(x)$ form an orthogonal basis in the Hilbert space of functions that are measurable and square integrable on $-\infty < x < +\infty$.

Now, we define the following elements in Hilbert space $\mathcal{H}_{2n}$

$$\tilde{h}_{m_1...m_n}^{m'_1...m'_n} = h_{m_1}(q_1)...h_{m_n}(q_n)h_{m'_1}(p_1)...h_{m'_n}(p_n), \qquad (17.2)$$
$$(m_1,...m_n, m'_1,...m'_n = 0,1,...),$$

and their transforms by the unitary transformation (16.1)

$$h_{m_1...m_n}^{m'_1...m'_n} = W\tilde{h}_{m_1...m_n}^{m'_1...m'_n}. \qquad (17.3)$$

The $h_{m_1...m_n}^{m'_1...m'_n}$ clearly depend on α and form an orthogonal basis in $\mathcal{H}_{2n}$. They are elements of Δ_{2n} and the operators P_j, Q_j, P'_j, Q'_j defined by (16.2), (16.3), (16.4) give, by virtue of (17.1):

$$\frac{1}{2}(\alpha^2 P_j^2 + Q_j^2)h_{m_1...m_n}^{m'_1...m'_n} = \left(m_j + \frac{1}{2}\right)h_{m_1...m_n}^{m'_1...m'_n};$$
$$(j = 1,...n). \quad (17.4)$$
$$\frac{1}{2}(\alpha^2 P_j'^2 + Q_j'^2)h_{m_1...m_n}^{m'_1...m'_n} = \left(m'_j + \frac{1}{2}\right)h_{m_1...m_n}^{m'_1...m'_n}.$$

Moreover, the only elements of $\mathcal{H}_{2n}$ that satisfy the preceding $2n$ relations are the elements $ah_{m_1...m_n}^{m'_1...m'_n}$, with a complex.

Let us prove the following preliminary proposition:

LEMMA: Take a Hilbert space $\mathcal{H}$, a hypermaximally symmetric operator A of domain D in $\mathcal{H}$, and a decomposition of $\mathcal{H}$ into a direct sum of two linearly closed manifolds $\mathcal{M}_1, \mathcal{M}_2$, invariant under the one-parameter group of unitary transformations $e^{i\tau A}$. The domain D is the direct sum of domains $D_1 = \mathcal{M}_1 \cap D$, $D_2 = \mathcal{M}_2 \cap D$; further, the restriction of A to D_κ, $(\kappa = 1, 2)$ is hypermaximally symmetric and maps D_κ into $\mathcal{M}_\kappa$.

This proposition is a very simple consequence of the theorems by Stone reviewed in section 2. For every $\psi \in \mathcal{H}$, denote its orthogonal projections onto $\mathcal{M}_1, \mathcal{M}_2$ by $E_1\psi, E_2\psi$, respectively.[37] They are orthogonal to each other and in fact $\psi = E_1\psi + E_2\psi$. If $\psi \in D$, we have (section 2, theorem C)

$$iA\psi = \lim_{\tau \to 0} \tau^{-1}[e^{i\tau A}\psi - \psi] = \lim_{\tau \to 0} \tau^{-1}\sum_{1,2}[e^{i\tau A}E_\kappa\psi - E_\kappa\psi]$$
$$= \sum_{1,2} \lim_{\tau \to 0} \tau^{-1}[e^{i\tau A}E_\kappa\psi - E_\kappa\psi].$$

For $\kappa = 1, 2$, the last limit exists and belongs to $\mathcal{M}_\kappa$. Then $E_\kappa\psi \in D$, $AE_\kappa\psi \in \mathcal{M}_\kappa$. It follows that every element of D is a sum of elements of $D_\kappa = \mathcal{M}_\kappa \cap D$ $(\kappa = 1, 2)$ and that $AD_\kappa \subset \mathcal{M}_k$. Thus, the lemma is proven.

On the other hand, recall that the infinitesimal transformations $X[f]$ in the subgroup L of Γ are obtained by taking f in the set $\mathcal{F}_L$ of polynomials of degree 0, 1 or 2 in the p_j, q_j, ($\mathcal{F}_L \subset \mathcal{F}_\Gamma$). The one-parameter subgroups of the representation $\mathcal{R}^{(\alpha)}$ of L are then transformations $e^{i\tau H^{(\alpha)}[f]}$ for $f \in \mathcal{F}_L$.

Given this, and for fixed α, let $\mathcal{M}_1$ be a linearly closed manifold in $\mathcal{H}_{2n}$, invariant under the transformations $U_\gamma^{(\alpha)}$, ($\gamma \in L$) in the representation $\mathcal{R}^{(\alpha)}$. Let $\mathcal{M}_2$ be the linearly closed manifold of elements in $\mathcal{H}_{2n}$ that are orthogonal to $\mathcal{M}_1$; the manifold $\mathcal{M}_2$ is also invariant. The orthogonal projection operator onto $\mathcal{M}_\kappa$ will again be denoted by E_κ ($\kappa = 1, 2$).

The element $h_{m_1...m_n}^{m'_1...m'_n}$ of $\mathcal{H}_{2n}$ enjoys the following properties. It belongs to the domains of the operators $H^{(\alpha)}[p_j] = \alpha P_j$, $H^{(\alpha)}[q_j] = \alpha Q_j$, and so does its transform by each of these operators. We have by (17.4)

$$\frac{1}{2\alpha^2}\left\{\alpha^2(H^{(\alpha)}[p_j])^2 + (H^{(\alpha)}[q_j])^2\right\}h_{m_1...m_n}^{m'_1...m'_n} = \left(m_j + \frac{1}{2}\right)h_{m_1...m_n}^{m'_1...m'_n}.$$

It also belongs to the domain of the operators (see 16.9)

$$H^{(\alpha)}[p_j^2] = \alpha(P_j^2 - P'^2_j), H^{(\alpha)}[q_j^2] = \alpha(Q_j^2 - Q'^2_j)$$

and according to (17.4) we have

$$\frac{1}{2\alpha}\left\{\alpha^2 H^{(\alpha)}[p_j^2] + H^{(\alpha)}[q_j^2]\right\}h_{m_1...m_n}^{m'_1...m'_n} = (m_j - m'_j)h_{m_1...m_n}^{m'_1...m'_n}.$$

Furthermore, every element of $\mathcal{H}_{2n}$ satisfying these properties is of the form $ah_{m_1...m_n}^{m'_1...m'_n}$ with a complex. Now, by the lemma showed above, $E_\kappa h_{m_1...m_n}^{m'_1...m'_n}$, ($\kappa = 1, 2$) satisfies these properties. Thus we have

$$E_\kappa h_{m_1...m_n}^{m'_1...m'_n} = a_\kappa h_{m_1...m_n}^{m'_1...m'_n}, \quad a_1 + a_2 = 1.$$

Since the manifolds $\mathcal{M}_1$ and $\mathcal{M}_2$ are orthogonal, one of the numbers a_1, a_2 must be zero. Therefore, *for every linearly closed manifold $\mathcal{M}_1$ of $\mathcal{H}_{2n}$ invariant under the $U_\gamma^{(\alpha)}$, ($\gamma \in L$), there exists a collection of elements $h_{m_1...m_n}^{m'_1...m'_n}$ spanning $\mathcal{M}_1$.*. The elements $h_{m_1...m_n}^{m'_1...m'_n}$ that are not in this collection span the manifold $\mathcal{M}_2$, orthogonal to $\mathcal{M}_1$.

Using (17.1) and the formulas of the previous section, one easily verifies that

$$i\alpha P_j h_{m_1...m_n}^{m'_1...m'_n} = m_j h_{m_1-\delta_{1j}...m_n-\delta_{nj}}^{m'_1...m'_n} - \frac{1}{2} h_{m_1+\delta_{1j}...m_n+\delta_{nj}}^{m'_1...m'_n} \quad (17.5)$$

$$i\alpha P'_j h_{m_1...m_n}^{m'_1...m'_n} = m'_j h_{m_1...m_n}^{m'_1-\delta_{1j}...m'_n-\delta_{nj}} - \frac{1}{2} h_{m_1...m_n}^{m'_1+\delta_{1j}...m'_n+\delta_{nj}}$$

Let $\mathcal{M}_1$ be a linearly closed manifold in $\mathcal{H}_{2n}$ invariant under $U_\gamma^{(\alpha)}$, $(\gamma \in L)$, and let $h_{m_1...m_n}^{m'_1...m'_n} \in \mathcal{M}_1$. By our lemma, $H^{(\alpha)}[p_j] h_{m_1...m_n}^{m'_1...m'_n} = \alpha P_j h_{m_1...m_n}^{m'_1...m'_n}$ is also in $\mathcal{M}_1$. By (17.5), we thus have $h_{m_1...m_n}^{m'_1...m'_n} \in \mathcal{M}_1$ regardless of the values of $m_1 \ldots m_n$.[t38] Analogously, the element

$$H^{(\alpha)}[p_j p_k] h_{m_1...m_n}^{m'_1...m'_n} = \alpha(P_j P_k - P'_j P'_k) h_{m_1...m_n}^{m'_1...m'_n}$$

belongs to $\mathcal{M}_1$. Then, one easily finds that $\mathcal{M}_1$ contains all the $h_{m_1...m_n}^{m'_1...m'_n}$ for which the sum $m_1 + \ldots + m_n$ has the same parity as $m'_1 + \ldots + m'_n$.[t39]

Consequently, there are only two linearly closed manifolds of $\mathcal{H}_{2n}$ that can be invariant under the $U_\gamma^{(\alpha)}$, $(\gamma \in L)$; they are the manifold $\mathcal{M}_+^{(\alpha)}$ spanned by the set of $h_{m_1...m_n}^{m'_1...m'_n}$ for which the sum $m'_1 + \ldots + m'_n$ is even, and the manifold $\mathcal{M}_-^{(\alpha)}$ spanned by the set of $h_{m_1...m_n}^{m'_1...m'_n}$ for which the sum is odd.

These manifolds are orthogonal and span the entire space $\mathcal{H}_{2n}$. We note that they depend on the value α under consideration.

We know that the Hermite function $h_m(x)$ is even or odd in x depending on whether the index m is even or not. By (17.2) and (17.3), the manifold $W^{-1}\mathcal{M}_+^{(\alpha)}$ thus consists of elements $\varphi(p_1, \ldots p_n, q_1, \ldots q_n)$ of $\mathcal{H}_{2n}$ for which

$$\varphi(-p_1, \ldots -p_n, q_1, \ldots q_n) = \varphi(p_1, \ldots p_n, q_1, \ldots q_n)$$

whereas $W^{-1}\mathcal{M}_-^{(\alpha)}$ contains the elements for which

$$\varphi(-p_1, \ldots -p_n, q_1, \ldots q_n) = -\varphi(p_1, \ldots p_n, q_1, \ldots q_n).$$

These equations must obviously be understood to hold up to a set of measure zero.

We have established the second part of the following statement:

THEOREM: *For all real nonzero α, the linearly closed manifolds $\mathcal{M}_+^{(\alpha)}$, $\mathcal{M}_-^{(\alpha)}$ of $\mathcal{H}_{2n}$ are invariant under the representation $\mathcal{R}^{(\alpha)}$ of the subgroup L of Γ, and are the only ones that enjoy this property.*

It remains to show that $\mathcal{M}_+^{(\alpha)}$, $\mathcal{M}_-^{(\alpha)}$ are indeed invariant. In the representation $\mathcal{R}^{(\alpha)}$, the one-parameter subgroups of L are represented by the unitary groups of transformation $e^{i\tau H^{(\alpha)}[f]}$, where f runs over the set $\mathcal{F}_L$ of polynomials of degree 0, 1 and 2 in p_j, q_j. For these elements f, the domain of definition of $H^{(\alpha)}[f]$ contains the domain Δ_{2n} defined in section 16. Since $\mathcal{D}_{2n} \in \Delta_{2n}$, the restriction of $H^{(\alpha)}[f]$ to Δ_{2n} is essentially hypermaximal. The domain Δ_{2n} is clearly the direct sum of the domains $\Delta_{2n} \cap \mathcal{M}_+^{(\alpha)} = \Delta_{2n}^+$ and $\Delta_{2n} \cap \mathcal{M}_-^{(\alpha)} = \Delta_{2n}^-$, that is for all $\varphi \in \Delta_{2n}$ one has $\varphi = \varphi_+ + \varphi_-$ with $\varphi_+ \in \Delta_{2n}^+$ and $\varphi_- \in \Delta_{2n}^-$. For $f \in \mathcal{F}_L$, we check that $H^{(\alpha)}[f]\Delta_{2n}^+ \in \Delta_{2n}^+$

and $H^{(\alpha)}[f]\Delta_{2n}^- \in \Delta_{2n}^-$. Using (16.9) we indeed find that for $f \in \mathcal{F}_L$, $H^{(\alpha)}[f]$ expressed in terms of the operators P_j, Q_j, P_j', Q_j' contains terms of degree 1 and 2 in the P_j, Q_j and only terms of degree 2 in the P_j', Q_j'. Recalling (16.2), (16.3) and (16.4) one immediately verifies our statement.

Now, we know (lemma I of section 13) that the domains $\{H^{(\alpha)}[f] \pm i\}\Delta_{2n}$ are dense in $\mathcal{H}_{2n}$. It follows that $\{H^{(\alpha)}[f] \pm i\}\Delta_{2n}^+$ and $\{H^{(\alpha)}[f] \pm i\}\Delta_{2n}^-$ are dense in $\mathcal{M}_+^{(\alpha)}$ and $\mathcal{M}_-^{(\alpha)}$, respectively. Thus, the restrictions of $H^{(\alpha)}[f]$ to Δ_{2n}^+ and Δ_{2n}^- are essentially hypermaximal; the domain of definition of $H^{(\alpha)}[f]$ is the direct sum of its intersections with $\mathcal{M}_+^{(\alpha)}$, $\mathcal{M}_-^{(\alpha)}$ and it maps them to $\mathcal{M}_+^{(\alpha)}$ and $\mathcal{M}_-^{(\alpha)}$, respectively. It follows that $e^{i\tau H^{(\alpha)}[f]}$ leaves these manifolds invariant.

The manifolds $\mathcal{M}_+^{(\alpha)}$, $\mathcal{M}_-^{(\alpha)}$ are invariant for the transformations that represent the one-parameter subgroups of L in $\mathcal{R}^{(\alpha)}$. We will show that L is an analytic, connected group. Since every analytic, connected group is generated by its continuous one-parameter subgroups (Chevalley [5], corollary 2, p. 118 and theorem 1, p. 35), the invariance of $\mathcal{M}_+^{(\alpha)}$, $\mathcal{M}_-^{(\alpha)}$ under the representation $\mathcal{R}^{(\alpha)}$ of L follows. The invariant subgroup T of L (see section 4) is manifestly connected. The quotient group L/T, being analytically isomorphic to the group of real symplectic matrices of rank $2n$, does not admit any trivial invariant continuous subgroups; indeed, Dickson has shown that the only nontrivial invariant subgroup of the real symplectic group is its center, consisting of the unit matrix and its product by -1 (see Dieudonné [6], p. 12). It follows that L/T is connected (Chevalley [5], proposition 1, p. 35). The groups T and L/T are connected, so L is connected as well (ibid, proposition 2, p. 36). Our theorem is then completely established.

18 Irreducibility of the Representation $\mathcal{R}^{(\alpha)}$ of the Group Γ for $\alpha \neq 0$

According to the previous section, the only linearly closed manifolds in $\mathcal{H}_{2n}$ that are invariant under the transformations $U_\gamma^{(\alpha)}$, ($\alpha \neq 0$, $\gamma \in L$) are $\mathcal{M}_+^{(\alpha)}$ and $\mathcal{M}_-^{(\alpha)}$. Thus, to prove the irreducibility of $\mathcal{R}^{(\alpha)}$, ($\alpha \neq 0$) as a representation of the entire group Γ, it suffices to show that certain transformations $U_\gamma^{(\alpha)}$ ($\gamma \in \Gamma$) do not leave invariant the manifolds $\mathcal{M}_+^{(\alpha)}$ and $\mathcal{M}_-^{(\alpha)}$. We will take the transformations $e^{i\tau H^{(\alpha)}[f_0]}$ with $f_0 = e^{-q_1^2}$. One easily sees that the domain of definition of $H^{(\alpha)}[f_0]$ contains the domain Δ_{2n}. For $\varphi \in \Delta_{2n}$,

(16.9) is valid and gives

$$\frac{1}{\alpha} H^{(\alpha)}[f_0]\varphi = [1 + 2Q_1'(Q_1 + Q_1')]e^{-(Q_1+Q_1')^2}\varphi \; ;$$

yielding

$$\frac{1}{\alpha} W^{-1} H^{(\alpha)}[f_0]\varphi = (1 + 2p_1q_1 + 2p_1^2)e^{-(p_1+q_1)^2} W^{-1}\varphi \; . \tag{18.1}$$

If the manifold $\mathcal{M}_+^{(\alpha)}$ were invariant under the transformations $e^{i\tau H^{(\alpha)}[f_0]}$ for $\varphi \in \mathcal{M}_+^{(\alpha)} \cap \Delta_{2n}$, we would have $H^{(\alpha)}[f_0]\varphi \in \mathcal{M}_+^{(\alpha)}$. Then, the function $W^{-1} H^{(\alpha)}[f_0]\varphi$ given by (18.1) would be even in $p_1, \ldots p_n$, which would then also be the case for the function $W^{-1}\varphi$. Now (18.1) shows that this is not true. Thus, the manifold $\mathcal{M}_+^{(\alpha)}$ is not invariant and the manifold $\mathcal{M}_-^{(\alpha)}$ cannot be invariant either. We have therefore established the following result:

THEOREM: *For $\alpha \neq 0$, the representation $\mathcal{R}^{(\alpha)}$ of the group Γ is irreducible.*

In section 10, our search for manifolds that are invariant under the representation $\mathcal{R}^{(\alpha)}$ was based on a slightly more precise fact that we will show presently: the existence of a countable sequence $\gamma_1, \ldots \gamma_k, \ldots$ in Γ, such that the space $\mathcal{H}_{2n}$ is irreducible for the set of transformations $U_{\gamma_1}^{(\alpha)}, \ldots U_{\gamma_k}^{(\alpha)}, \ldots$ whatever the value of $\alpha \neq 0$ may be. Let us consider the set of unitary transformations $e^{i\tau H^{(\alpha)}[f]}$ for τ rational and $f = p_j, q_j, p_j^2, q_j^2, p_j p_l, e^{-q_1^2}$ ($j, l = 1, \ldots n$); this set constitutes a countable sequence of transformations $U_{\gamma_1}^{(\alpha)}, \ldots U_{\gamma_k}^{(\alpha)}, \ldots$ where the $\gamma_k \in \Gamma$ are independent of α. By strong continuity in τ, a linear closed manifold in $\mathcal{H}_{2n}$, invariant under transformation in this sequence, is invariant under the transformations $e^{i\tau H^{(\alpha)}[f]}$ for all real τ and the f given above. The reasoning of section 17 and of the present section uses only these transformations for showing irreducibility of the space $\mathcal{H}_{2n}$. Thus, for every $\alpha \neq 0$, $\mathcal{H}_{2n}$ is irreducible under the set of transformations $U_{\gamma_1}^{(\alpha)}, \ldots U_{\gamma_k}^{(\alpha)}, \ldots$. The series $\gamma_1, \ldots \gamma_k, \ldots$ of elements of Γ has the property we used in section 10.

19 Irreducibility of the representation $\mathcal{R}^{(0)}$

The representation $\mathcal{R}^{(0)}$ differs strongly, in its structure, from the representations we have studied so far, mainly because the operator $\mathcal{H}^{(0)}[1]$ is zero and therefore the $2n$ operators $\mathcal{H}^{(0)}[p_j]$, $\mathcal{H}^{(0)}[q_j]$ commute. In contrast with $\mathcal{R}^{(\alpha)}$ for $\alpha \neq 0$, it has the following property:

THEOREM: *The representation $\mathcal{R}^{(0)}$ of the subgroup L of Γ is irreducible.*

To prove this theorem, consider the commuting operators $\mathcal{H}^{(0)}[p_j]$, $\mathcal{H}^{(0)}[q_j]$. By (12.1), they have the very simple form[t40]

$$\mathcal{H}^{(0)}[p_j] = \frac{1}{i}\frac{\partial}{\partial q_j}, \quad \mathcal{H}^{(0)}[q_j] = -\frac{1}{i}\frac{\partial}{\partial p_j}.$$

We obtain their spectral representation by the Fourier transform[t41]

$$V\varphi(p_1,\ldots p_n, q_1 \ldots q_n) = \qquad (19.1)$$
$$\left(\frac{1}{2\pi}\right)^n \int e^{i\sum_j(\tilde{p}_j q_j - p_j \tilde{q}_j)} \varphi(\tilde{p}_1,\ldots \tilde{p}_n, \tilde{q}_1 \ldots \tilde{q}_n) d\tilde{p}_1 \ldots d\tilde{p}_n\, d\tilde{q}_1 \ldots d\tilde{q}_n.$$

The operators

$$B'_j = V^{-1} H^{(0)}[p_j] V, \quad B_j = V^{-1} H^{(0)}[q_j] V$$

are operators of multiplication by p_j, q_j, respectively. For every set $\mathcal{E}$ of the space (p, q), we denote by $\mathcal{N}'(\mathcal{E})$ the linearly closed manifold composed of elements $\varphi(p, q)$ of $\mathcal{H}_{2n}$ for which $\varphi(p, q) = 0$ outside $\mathcal{E}$, except possibly on a set of measure zero. The manifolds $\mathcal{N}'(\mathcal{E})$, $\mathcal{N}'(\mathcal{E}')$ are identical when the sets $\mathcal{E}$, $\mathcal{E}'$ differ by a set of measure zero, and only then. It is well known that the manifolds $\mathcal{N}'(\mathcal{E})$ are the only linearly closed manifolds that are invariant under the set of transformations

$$e^{i\tau B_j}, \quad e^{i\tau B'_j}, \quad (j = 1,\ldots n; -\infty < \tau < +\infty).$$

Consequently, *the manifolds* $\mathcal{N}(\mathcal{E}) \equiv V\mathcal{N}'(\mathcal{E})$ *are the only linearly closed manifolds of* $\mathcal{H}_{2n}$ *that are invariant under the set of transformations*

$$e^{i\tau H^{(0)}[p_j]}, \quad e^{i\tau H^{(0)}[q_j]}, \qquad (19.2)$$

that is, under the representation $\mathcal{R}^{(0)}$ *of the subgroup* T *of* Γ (see section 4).

With this is mind, let us consider a homogeneous transformation γ of the subgroup L, given by the equations (see section 4)

$$p'_j = \sum_k (a_{jk} p_k + b_{jk} q_k), \quad q'_j = \sum_k (c_{jk} p_k + d_{jk} q_k), \qquad (19.3)$$

in the space (p, q), where the matrix

$$\begin{pmatrix} a_{jk} & b_{jk} \\ c_{jk} & d_{jk} \end{pmatrix}$$

is symplectic. If one subjects $\tilde{p}_j$, $\tilde{q}_j$ to the same transformation (19.3), the bilinear form $\sum_j (p_j \tilde{q}_j - \tilde{p}_j q_j)$ is invariant. By expression (19.1) for the transformation V, it follows that $V U^{(0)}_\gamma = U^{(0)}_\gamma V$, where $U^{(0)}_\gamma$ denotes the unitary transformation corresponding to γ in the representation $\mathcal{R}^{(0)}$, as usual (section 8). On the other hand, denoting by $\gamma\mathcal{E}$ the transform by γ of the measurable set $\mathcal{E}$ in the space (p,q), we immediately find $U^{(0)}_\gamma \mathcal{N}'(\mathcal{E}) = \mathcal{N}'(\gamma\mathcal{E})$. Thus

$$U^{(0)}_\gamma \mathcal{N}(\mathcal{E}) = U^{(0)}_\gamma V\mathcal{N}'(\mathcal{E}) = VU^{(0)}_\gamma \mathcal{N}'(\mathcal{E}) = V\mathcal{N}'(\gamma\mathcal{E}) = \mathcal{N}(\gamma\mathcal{E}) .$$

If there exists a linearly closed manifold in $\mathcal{H}_{2n}$ which is invariant under the representation $\mathcal{R}^{(0)}$ of the subgroup L, it is in particular invariant under the transformations (19.2). Consequently, it is a manifold $\mathcal{N}(\mathcal{E})$, where $\mathcal{E}$ is a measurable set in the space (p,q). Since it must be invariant also under the $U^{(0)}_\gamma$, with γ of type (19.3), we must have that, for every transformation of this type, $\mathcal{N}(\mathcal{E}) = \mathcal{N}(\gamma\mathcal{E})$. Thus, for every transformation γ of type (19.3) the sets $\mathcal{E}$ and $\gamma\mathcal{E}$ can only differ by a set of measure zero. Let us therefore consider the additive and absolutely continuous set function $F(E) = m(\mathcal{E} \cap E)$ on the space (p,q), for every measurable and bounded set E; m denotes the Lebesgue measure. As is known, at every point ω of the space (p,q) which does not belong to an exceptional set E_0 of measure zero, the function $F(E)$ admits the derivative 1 or 0 depending on whether ω is in $\mathcal{E}$ or not, and this derivative does not depend on the regular family of sets we used for defining it (see for example La Vallé Poussin [16], ch. IV and V, especially p. 64 for the definition of regular families, and the theorem of section 64, p. 76). On the other hand, by the property of $\mathcal{E}$ we just pointed out, we have $F(E) = F(\gamma E)$ for the transformations γ of type (19.3). These transformations indeed conserve the measure. Since they transform one regular family of sets into another, the derivative of $F(E)$ will have the same value at all points that are transforms of a given point, provided they are not in E_0. Given two points of the space (p,q) distinct from the origin $p_1 = \ldots = p_n = q_1 = \ldots = q_n = 0$, one of them is always the transform of the other under a transformation of type (19.3). Thus, the derivative of $F(E)$ is either 0 almost everywhere or 1 almost everywhere. We find that the set $\mathcal{E}$ is either of measure zero or differs from the space (p,q) only by a set of measure zero, and the manifold $\mathcal{N}(\mathcal{E})$ reduces to the zero element, or constitutes the entire space $\mathcal{H}_{2n}$. Thus our theorem is proven.

A brief outline of the results in this chapter is given in Van Hove [32].

CHAPTER VI

Comparison Between Quantum Mechanical and Classical Operators

After some general remarks on the relations between classical and quantum descriptions of a dynamical system, we point out some properties of the family of operators in Quantum Mechanics, and we compare this family to the family of operators $H^{(\alpha)}[f]$ that we defined in the context of Classical Mechanics. We then establish that a bijective correspondence between classical and quantum quantities, in the guise of an isomorphism between Lie algebras, exists for quantities represented by polynomials of degree 0, 1 or 2 in the variables $p_1 \ldots p_n, q_1 \ldots q_n$; it cannot be extended without losing properties essential to all classical quantities.

20 The Problem of Relations Between Quantum and Classical Descriptions of a System

In the preceding chapters we constructed and studied certain unitary representations of the group Γ in Hilbert space. We showed how the group Γ is naturally associated with a dynamical system in Classical Mechanics; physical quantities associated with the system (which one can identify with first integrals) are in correspondence with the infinitesimal transformations in Γ (section 1). Using unitary representations of Γ as intermediaries, the quantities of the system were put in correspondence with some hypermaximally symmetric operators on Hilbert space. This situation is quite analogous to the description of a dynamical system in Quantum Mechanics: there, too, a hypermaximally symmetric operator corresponds to every quantity associated with the system.

Assume that a dynamical system is described in the framework of Classical Mechanics on the one hand, and in Quantum Mechanics on the other. We encounter the oft-discussed problem of relations between the classical and quantum descriptions of the system. This problem has two aspects.

On one hand, it concerns proving that the classical description is a first approximation to the quantum description, the Planck constant h being considered negligible (Bohr's correspondence principle). The classical theory must appear as an asymptotic limit of the quantum theory when h tends to zero. Some well-known arguments show that formally this is indeed the case. However, one must realize that the mathematical nature of taking this limit is ill understood, and some questions appear that would perhaps deserve to be examined more closely. We do not approach those questions here.[t42]

On the other hand, observed phenomena that are described to sufficient approximation by Classical Mechanics are clearly much more immediate and easier to understand than purely quantum phenomena.[t43] The description of classical phenomena preceded the description of quantum phenomena, and the former had attained a high degree of perfection long before the birth of the latter. Thus, it is natural to ask ourselves which part of our understanding of the classical description of systems we can cull to put together the less well-known quantum description. This is the problem of quantization of a system described by Classical Mechanics, or the problem of passage from Classical Mechanics to Quantum Mechanics. For systems with a finite number of degrees of freedom, this problem has been the subject of numerous discussions during the period when de Broglie, Heisenberg, Schrödinger, Born and Jordan created Quantum Mechanics in its present form. Shortly thereafter, the problem reappeared in the context of systems with an infinite number of degrees of freedom (field quantization). Since then, many works have been dedicated to this line of study. The most original contribution, due to Feynman[8], is only a few years old.

Our goal is not to discuss in detail the two aspects of the problem of relations between classical and quantum descriptions of a system. We merely want to exhibit this problem from a new point of view. As we emphasized above, in the same way that Quantum Mechanics associates a Hilbert space operator with every quantity of a system, we have associated such operators to every quantity in the classical description of a system. Thus, it is natural to compare the operators corresponding to various quantities in the classical and quantum descriptions. We will make this comparison in the following sections.

21 Characterization of Operators in Quantum Mechanics

Consider the description in Quantum Mechanics of a dynamical system with n degrees of freedom, that we assume devoid of any natural property that is exclusively quantum in nature (that is, without classical analog; spin is one example). Physical quantities associated with the system are represented by hypermaximally symmetric operators on a separable Hilbert space $\mathcal{H}$. Certain systems of $2n$ quantities, pairwise canonically conjugate, play a fundamental role in the study of the system. They are represented by operators $P_1, \ldots P_n$, $Q_1, \ldots Q_n$ enjoying the following properties:

(a) The one-parameter groups of unitary transformations $e^{i\sigma P_j}$, $e^{i\tau Q_j}$, $(j = 1, \ldots n)$ satisfy the following commutation relations

$$e^{i\sigma P_j} e^{i\sigma' P_k} = e^{i\sigma' P_k} e^{i\sigma P_j}, \quad e^{i\tau Q_j} e^{i\tau' Q_k} = e^{i\tau' Q_k} e^{i\tau Q_j}$$
$$e^{i\sigma P_j} e^{i\tau Q_k} = e^{i\hbar\sigma\tau \delta_{jk}} e^{i\tau Q_k} e^{i\sigma P_j} \quad (j, k = 1, \ldots n), \tag{21.1}$$

where $\hbar = h/2\pi$ is Planck's constant divided by 2π and δ_{jk} the Kronecker symbol.

(b) The Hilbert space $\mathcal{H}$ is irreducible under the set of transformations

$$e^{i\sigma P_j}, \quad e^{i\sigma Q_j}, \quad (j = 1, \ldots n; \; -\infty < \sigma < +\infty)$$

By a theorem of Stone [29], properties (a) and (b) determine the operators P_j, Q_j up to a unitary transformation. Indeed, Stone proved the following theorem: given hypermaximally symmetric operators satisfying properties (a) and (b), there exists a bijective map S, linear and conserving the scalar product, from $\mathcal{H}$ to the Hilbert space $\mathcal{H}_n$ of complex-valued functions $g(x_1, \ldots x_n)$, measurable and square integrable in the n-dimensional Euclidean space $(x_1, \ldots x_n)$ with volume element $dx_1 \ldots dx_n$, such that

$$SP_j S^{-1} g(x_1, \ldots x_n) = \frac{\hbar}{i} \frac{\partial}{\partial x_j} g(x_1, \ldots x_n),$$
$$SQ_j S^{-1} g(x_1, \ldots x_n) = x_j g(x_1, \ldots x_n), \quad (j = 1, \ldots n). \tag{21.2}$$

This result by Stone was completed by von Neumann [22], who showed that if property (a) is satisfied without (b) being satisfied, the Hilbert space $\mathcal{H}$ is the direct sum of a finite or countably infinite number of subspaces invariant and irreducible under the operators $e^{i\sigma P_j}$, $e^{i\sigma Q_j}$. The theorem of Stone applies to each of these subspaces.

It is known that equations (21.1) lead to the well-known commutation relations

$$P_j P_k - P_k P_j = 0, \; Q_j Q_k - Q_k Q_j = 0, \; P_j Q_k - Q_k P_j = \frac{\hbar}{i} \delta_{jk}. \tag{21.3}$$

According to (21.2), these relations are valid on a dense domain of the space $\mathcal{H}$.

Now recall, following von Neumann[20], some notions concerning the algebra of bounded operators in the Hilbert space $\mathcal{H}$. This name is used for the family of bounded operators that contains products and linear combinations with complex coefficients of any two members of the family. An algebra of bounded operators is called self-adjoint when it contains the adjoint of each of its operators. A self-adjoint algebra of bounded operators that is closed under strong convergence and contains the identity transformation I on $\mathcal{H}$ is called a ring of bounded operators. It can be shown that such a ring is also closed in the weak topology. If $\mathcal{A}$ is a family of bounded operators on $\mathcal{H}$, the family of of bounded operators that commute with every element of $\mathcal{A}$ constitutes a ring. This ring is called the commutator of $\mathcal{A}$ and is denoted by $\mathcal{A}'$. The ring $\mathcal{A}'' = (\mathcal{A}')'$ is the smallest ring containing the elements of $\mathcal{A}$. It is also the strong and weak closure of the self-adjoint algebra generated by I and the elements of $\mathcal{A}$. We call $\mathcal{A}''$ the ring generated by $\mathcal{A}$. We clearly have $\mathcal{A}'' = \mathcal{A}$ when $\mathcal{A}$ is a ring. For proofs of these statements, we refer to the work of von Neumann cited previously.

Let us return to the operators P_j, Q_j considered above. We immediately obtain the following proposition: *if the property (b) is satisfied, the operators $e^{i\sigma P_j}$, $e^{i\sigma Q_j}$, $(j = 1,\ldots n; -\infty < \sigma < +\infty)$ generate the ring of all bounded operators on $\mathcal{H}$.* Indeed, denote by $\mathcal{A}$ the ring generated by these operators. If $A \in \mathcal{A}'$, the bounded symmetric operators $A+A^*$, $i(A-A^*)$ belong to $\mathcal{A}'$ and their spectral manifolds are invariant under the operators in $\mathcal{A}$. According to property *(b)*, the space $\mathcal{H}$ is irreducible under the operators of $\mathcal{A}$. Thus, every $A \in \mathcal{A}'$ is of the form $A = cI$, c complex, and $\mathcal{A} = \mathcal{A}''$ is the ring of all bounded operators on Hilbert space.

We know that quantities attached to our dynamical system are represented by the set of hypermaximally symmetric operators on $\mathcal{H}$. Noting that each one of these is a function of a symmetric operator bounded by $\mathcal{H}$, we obtain the desired conclusion. *The family $\mathcal{Q}$ of operators that represent quantities associated to a dynamical system in Quantum Mechanics can be generated as follows; one chooses $2n$ operators P_j, Q_j, $(j = 1,\ldots n)$ with properties (a), (b). One constructs symmetric operators expressed by the finite sums*

$$\sum_{\sigma_j, \tau_j} a(\sigma_1,\ldots \sigma_n, \tau_1,\ldots \tau_n) e^{i \sum (\sigma_j P_j + \tau_j Q_j)} ; \qquad (21.4)$$

$$a(-\sigma_j, -\tau_j) = \bar{a}(\sigma_j, \tau_j) .$$

They constitute a family $\mathcal{B}$, where one takes the closure $\mathcal{Q}_b$ under strong convergence. $\mathcal{Q}$ is identical to the family of real functions of operators of $\mathcal{Q}_b$. We note that $\mathcal{Q}_b$ is the family of bounded operators of $\mathcal{Q}$. One must also note that in the above construction of $\mathcal{Q}$, the property of irreducibility *(b)* plays no role. If it is not satisfied, the resulting family $\mathcal{Q}$ is the direct sum of a finite or countably infinite number of isomorphic families for which the property *is* satisfied.[t44]

The family $\mathcal{Q}$ can be found from the family $\mathcal{B}$ of operators (21.4) by taking a limit. *Thus, one can formally give an expression of the following type to each element of $\mathcal{Q}$:*

$$\int a(\sigma_1,\ldots \sigma_n, \tau_1,\ldots \tau_n) e^{i \sum (\sigma_j P_j + \tau_j Q_j)} d\sigma_1 \ldots d\sigma_n\, d\tau_1 \ldots d\tau_n , \qquad (21.5)$$

with $a(-\sigma_j, -\tau_j) = \bar{a}(\sigma_j, \tau_j)$. The integral should be interpreted as the limit of sums (21.4) in a very general sense. The formal expression (21.5), which was first proposed by Weyl [34], will be easy to compare to the operators that we have introduced in the Classical Mechanics of the system.

22 Comparison of the Operators $H^{(\alpha)}[f]$ with the Operators of Quantum Mechanics

Let us return to the operators $H^{(\alpha)}[f]$ that we studied in previous chapters. These operators are associated to elements f of $\mathcal{F}_\Gamma$, which as we know represent the quantities associated with a system of n degrees of freedom in Classical Mechanics. Again using the notation of section 16, we take as in (16.5)

$$\frac{1}{\alpha} H^{(\alpha)}[p_j] = P_j \;, \quad \frac{1}{\alpha} H^{(\alpha)}[q_j] = Q_j \;. \tag{22.1}$$

If one identifies α with $2\pi/h$, where h denotes Planck's constant, the commutation relations obtained in section 16 for the operators (22.1) coincide with the relations (21.3) and the groups of transformation $e^{i\sigma P_j}$, $e^{i\sigma Q_j}$ satisfy property *(a)* of the preceding paragraph. However, property *(b)* is not satisfied: the Hilbert space $\mathcal{H}_{2n}$ is not irreducible under the operators $e^{i\sigma P_j}$, $e^{i\sigma Q_j}$.

On the other hand, consider the operators P'_j, Q'_j defined in (16.4). They commute with the operators (22.1). For $\alpha = 2\pi/h$, they also have commutation relations of the form (21.3) and the groups $e^{i\sigma P'_j}$, $e^{i\sigma Q'_j}$ also satisfy the property *(a)*. Further, the space $\mathcal{H}_{2n}$ is irreducible under the set of operators $e^{i\sigma P_j}$, $e^{i\sigma Q_j}$, $e^{i\sigma P'_j}$, $e^{i\sigma Q'_j}$, $(j = 1,\ldots n;\ -\infty < \sigma < \infty)$. Let $\mathcal{A}$ be the ring of bounded operators on $\mathcal{H}_{2n}$ generated by the operators $e^{i\sigma P_j}$, $e^{i\sigma Q_j}$ $(j = 1,\ldots n;\ -\infty < \sigma < \infty)$; let $\tilde{\mathcal{A}}$ be the ring generated by $e^{i\sigma P'_j}$, $e^{i\sigma Q'_j}$, $(j = 1,\ldots n;\ -\infty < \sigma < \infty)$. If $\alpha = 2\pi/h$, each of these rings is algebraically and topologically isomorphic to the ring generated by the operators of the quantum theory. On the other hand, by (16.9), the operators of the classical theory are expressed as functions of the two isomorphic rings $\mathcal{A}$ and $\tilde{\mathcal{A}}$.

Let us recall the definition given in (16.2), (16.3) and (16.4) for the operators P_j, Q_j, P'_j, Q'_j and take into account the results of the previous section. Using the unitary transformation W defined in (16.1), one sees that $W^{-1}\mathcal{A}W$ is the ring of bounded operators in $\mathcal{H}_{2n}$ acting on the variables $q_1,\ldots q_n$, whereas $W^{-1}\tilde{\mathcal{A}}W$ is the ring of operators acting on the variables $p_1,\ldots p_n$. We easily find that each of the two rings is the commutator of the other. Consequently, *each of the rings $\mathcal{A}$, $\tilde{\mathcal{A}}$ is the commutator of the other.*

By analogy with expression (21.4) for elements of the family $\mathcal{B}$, consider elements expressed by the finite sums

$$f(p,q) = \sum_{\sigma_j, \tau_j} a(\sigma_1, \ldots \sigma_n, \tau_1, \ldots \tau_n) \, e^{i \sum (\sigma_j P_j + \tau_j Q_j)} \, ; \qquad (22.2)$$

$$a(-\sigma_j, -\tau_j) = \bar{a}(\sigma_j, \tau_j) \, .$$

They belong to $\mathcal{F}_\Gamma$. The family they constitute in $\mathcal{F}_\Gamma$, being closed under Poisson brackets and linear combinations with real coefficients, is a Lie algebra. Let us apply (16.9) to an element of this family. We find

$$\frac{1}{\alpha} H^{(\alpha)}[f] = \qquad (22.3)$$

$$\sum_{\sigma_j, \tau_j} a(\sigma_1, \ldots \sigma_n, \tau_1, \ldots \tau_n) \, e^{i \sum [\sigma_j (P_j - P'_j) + \tau_j (Q_j + Q'_j)]} \, [1 + i \sum (\sigma_j P'_j - \tau_j Q'_j)] \, .$$

Now, compare this operator with the operator (21.4). It differs from the latter by replacing

$$e^{i \sum (\sigma_j P_j + \tau_j Q_j)}$$

by

$$e^{i \sum [\sigma_j (P_j - P'_j) + \tau_j (Q_j + Q'_j)]} \, [1 + i \sum (\sigma_j P'_j - \tau_j Q'_j)] = \qquad (22.4)$$

$$e^{i \sum (\sigma_j P_j + \tau_j Q_j)} A(\sigma_1, \ldots \sigma_n, \tau_1, \ldots \tau_n) \, .$$

with

$$A(\sigma_1, \ldots \sigma_n, \tau_1, \ldots \tau_n) = e^{i \sum_j (\tau_j Q'_j - \sigma_j P'_j)} [1 + i \sum (\sigma_j P'_j - \tau_j Q'_j)] \, . \qquad (22.5)$$

Thus, the quantum operators (21.4) and the classical operators (22.3) for $\alpha = 2\pi/h$ differ by the appearance in the classical operators of the factor $A(\sigma_j, \tau_j)$. We have to note a remarkable fact concerning the factor $A(\sigma_j, \tau_j)$; if we expand the operator A in powers of $P'_j \, Q'_j$, the term of degree zero is the identity transformation and there is no linear term in the $P'_j \, Q'_j$. It can be shown that, once we consider Planck's constant infinitely small (that is, α infinitely large) and study the passage from Quantum Mechanics to Classical Mechanics, we must consider the operators P'_j, Q'_j to be infinitely small. Then the factor A, which marks the difference between the two theories in our formulas, differs from the identity transformation only by some numbers that are infinitely small to second order. In our formalism, this is an indication of the convergence of the two theories toward each other when h tends to zero.

From (21.4) we were led to the formal expression (21.5) for all operators of Quantum Mechanics. Similarly, from (22.3),(22.4) and (22.5) for the operators of Classical Mechanics, we can write the formal expression

$$\frac{1}{\alpha}H^{(\alpha)}[f] = \int a(\sigma_1,\ldots\sigma_n,\tau_1,\ldots\tau_n)e^{i\sum(\sigma_j P_j+\tau_j Q_j)} \qquad (22.6)$$
$$A(\sigma_1,\ldots\sigma_n,\tau_1,\ldots\tau_n)\,d\sigma_1\ldots d\sigma_n\,d\tau_1\ldots d\tau_n$$

with $a(-\sigma_j,-\tau_j) = \bar{a}(\sigma_j,\tau_j)$. We note that $a(\sigma_j,\tau_j)$ is none other than the Fourier transform, which we assumed exists, of the element f of $\mathcal{F}_\Gamma$ with respect to the $2n$ variables p_j, q_j. Here again, the operator $A(\sigma_j,\tau_j)$ marks the difference between classical and quantum theories.

We can establish a correspondence between the quantities in Classical Mechanics and those of Quantum Mechanics by deciding that two quantities correspond to each other if they can be expressed by the same function $a(\sigma_j,\tau_j)$. This correspondence was introduced by Weyl [34].

23 Difficulties with a Unique Correspondence Between Classical and Quantum Quantities

As we have seen, the representation $\mathcal{R}^{(\alpha)}$ of the group Γ for $\alpha = 2\pi/h$ puts classical quantities, represented by the elements f of $\mathcal{F}_\Gamma$, in correspondence with the operators $K[f] = \frac{1}{\alpha}H^{(\alpha)}[f]$, $(\alpha = 2\pi/h)$ which present certain analogies with the operators of Quantum Mechanics. Furthermore, by corollary II of section 12, the operators $K[f]$ have the following property:

(q_1) If $f_1, f_2, a_1f_1 + a_2f_2 \in \mathcal{F}_\Gamma$ and a_1, a_2 real, we have

$$K[a_1f_1 + a_2f_2] = a_1K[f_1] + a_2K[f_2] \; ;$$

if $f_1, f_2, (f_1,f_2) \in \mathcal{F}_\Gamma$ we have

$$\frac{h}{2\pi}iK[(f_1,f_2)] = K[f_1]K[f_2] - K[f_2]K[f_1] \; .$$

These relations are valid on a domain where the operators are essentially hypermaximal.

In the quantization of dynamical systems described by Classical Mechanics, we associate the simplest quantities of the classical theory with operators having property (q_1), and we use this property to construct these operators as functions of the fundamental operators P_j, Q_j of the quantum theory (see for example Dirac [7] (p. 88-91). It is well known that by this procedure, one cannot associate a well-defined operator to every classical quantity such that

(q_1) is satisfied without exceptions. Products of operators give rise to ordering ambiguities that seem inevitable.[t45] In this context, we should alert the reader to a result of Groenewold [13] (p. 45) which proves that a unique correspondence $f \to K[f]$ between polynomials in the classical quantities p_j, q_j and polynomials in the quantum operators P_j, Q_j cannot, without exceptions, satisfy the relation

$$\frac{h}{2\pi} iK[(f_1, f_2)] = K[f_1]K[f_2] - K[f_2]K[f_1] .$$

In this state of affairs, the question arises whether there exists a unitary representation of the group Γ, or at least a unitary representation of the system of its infinitesimal transformations $X[f]$, ($f \in \mathcal{F}_\Gamma$), associating the simplest elements f to some hypermaximally symmetric operators that are identical to operators of Quantum Mechanics.[t46] The existence of such a representation is rendered improbable by the result of Groenewold. However, his result does not prove it is impossible, since Groenewold relies on classical quantities (such as $p_j q_j^2$, $p_j^2 q_j$) which do not belong to $\mathcal{F}_\Gamma$.[i] We will now give a rigorous proof of this impossibility. In the following section, we will show that on the other hand, a representation of the type mentioned above exists for the infinitesimal transformations in the subgroup L of Γ.

Let $\mathcal{H}_n$ be the Hilbert space of functions $\varphi(x_1, \ldots x_n)$ of real variables $x_1, \ldots x_n$, measurable and square integrable on n-dimensional Euclidean space, with scalar product

$$(\varphi, \psi) = \int \bar{\varphi}\psi \, dx_1 \ldots dx_n ,$$

where the integral ranges of the entire space. With α a real number, we consider the $2n$ hypermaximally symmetric operators

$$\begin{cases} \Pi_j \varphi(x_1, \ldots x_n) = \dfrac{1}{\alpha i} \dfrac{\partial}{\partial x_j} \varphi(x_1, \ldots x_n) \,; \\ R_j \varphi(x_1, \ldots x_n) = x_j \varphi(x_1, \ldots x_n) \,, \qquad (j = 1, \ldots n) \,. \end{cases} \quad (23.1)$$

---

[i] One can only demand that a classical quantity $f(p, q)$ which does not belong to $\mathcal{F}_\Gamma$ corresponds to a hypermaximally symmetric operator in the quantum theory. Indeed, if $f(p, q)$ does not belong to $\mathcal{F}_\Gamma$, this function taken as Hamiltonian gives rise to equations of motion for which certain regions of phase space (p, q) are sent to infinity after a finite time interval. This is the case, for example, for $f = p_j q_j^2$, $f = p_j^2 q_j$. In the quantum theory, on the other hand, f corresponds to a hypermaximally symmetric operator A, and time evolution of the system is represented by the transformation $e^{2\pi i t A/h}$ where the unitary character excludes loss to infinity after finite time.

For $\alpha = 2\pi/h$, they reduce to the Schrödinger operators (21.2). Further, given an infinitesimal transformation $X[f]$, $(f \in \mathcal{F}_\Gamma)$ of the group Γ, we denote by $\gamma_\tau^{(f)}$ the elements of the one-parameter subgroup generated by $X[f]$ in Γ.

Thus, we can state the following theorem:

THEOREM: *For real α, there does not exist a map $f \to A[f]$ of $\mathcal{F}_\Gamma$ in the set of hypermaximally symmetric operators of the Hilbert space $\mathcal{H}_n$ that simultaneously satisfies the following conditions (q'_1) and (q'_2):*

(q'_1) *There exists a domain $\mathcal{D}$ in $\mathcal{H}_n$ with the following properties:*[47]

- *it is common to all domains of definition of the operators $A[f]$, $(f \in \mathcal{F}_\Gamma)$*

- *every $A[f]$ and every $e^{i\alpha\tau A[f]}$ $(f \in \mathcal{F}_\Gamma, \tau$ real$)$ map $\mathcal{D}$ into itself.*

- *Let f_1 and $f_2 \in \mathcal{F}_\Gamma$, and a_1, a_2 real. The relations $a_1 f_1 + a_2 f_2 \in \mathcal{F}_\Gamma$ and $(f_1, f_2) \in \mathcal{F}_\Gamma$ give, respectively,*

$$\{a_1 A[f_1] + a_2 A[f_2]\}\varphi = A[a_1 f_1 + a_2 f_2]\varphi \,;$$

$$\{A[f_1]A[f_2] - A[f_2]A[f_1]\}\varphi = \frac{i}{\alpha}A[(f_1, f_2)]\varphi$$

for all $\varphi \in \mathcal{D}$.

- *Finally, the map $\gamma_\tau^{(f)} \to e^{i\alpha\tau A[f]}$ $(f \in \mathcal{F}_\Gamma, \tau$ real$)$ satisfies*

$$\gamma_\sigma^{(g')} = \gamma_\tau^{(f)}\gamma_\sigma^{(g)}\gamma_{-\tau}^{(f)} \to e^{i\alpha\sigma A[g']} = e^{i\alpha\tau A[f]}e^{i\alpha\sigma A[g]}e^{-i\alpha\tau A[f]}$$

for all $f, g, g' \in \mathcal{F}_\Gamma$ and σ, τ real.

(q'_2) *For $j = 1, \ldots n$, the operators $A[p_j]$, $A[q_j]$ are identical to the operators Π_j, R_j of (23.1), respectively.*

The last part of condition (q'_1) is equivalent to the following condition: if $X[g']$ is the transform in Γ of the infinitesimal transformation $X[g]$ by the inner automorphism $\gamma \to \gamma_\tau^{(f)}\gamma\gamma_{-\tau}^{(f)}$, the hypermaximally symmetric operators $A[g']$ and $e^{i\alpha\tau A[f]}A[g]e^{-i\alpha\tau A[f]}$ are identical. Thus, one sees that this condition concerns the representation in $\mathcal{H}_n$ of the adjoint group[j] of Γ', where Γ' is the subgroup of Γ spanned by the set of its one-parameter subgroups.

Given the irreducibility of $\mathcal{H}_n$ under the operators $e^{i\tau\Pi_j}$, $e^{i\tau R_j}$, the set $\mathcal{D}$ is dense in $\mathcal{H}_n$ provided it contains a nonzero element, which we assume.

---

[j] The adjoint group is a group of maps from the set $\mathcal{F}_\Gamma$ onto itself. These maps respect Poisson brackets and linear combination with real coefficients whenever they are defined. The elements g and g' above are functions of points ω in the space (p, q). They are related by $g'(\omega) = g(\gamma_{-\tau}^{(f)}\omega)$. For a finite-dimensional Lie group the adjoint group is none other than the linear adjoint representation, defined in Chevalley [5] (p. 123), for example.

One immediately sees that for $\alpha = 2\pi/h$ the condition (q_1') is a more restrictive formulation of another property (q_1) given below. It is possible that the various parts of (q_1') are not entirely independent; however, we will not concern ourselves with questions arising from this possibility.

We note that the theorem stated above no longer holds if one neglects condition (q_2'). Indeed, the results of chapter IV, section 18, show that the condition (q_1') is satisfied if one replaces the space $\mathcal{H}_n$ by the space $\mathcal{H}_{2n}$ and if one takes

$$A[f] \equiv \frac{1}{\alpha} H^{(\alpha)}[f], \quad \mathcal{D} \equiv \mathcal{D}_{2n}.$$

Now, we prove this by *reductio ad absurdum*.[t48] Assume there exists a map $f \to A[f]$ that satisfies the conditions (q_1') and (q_2'). The domain $\mathcal{D}$ belongs to the domain of definition of the operators Π_j, R_j, and these operators map $\mathcal{D}$ into itself. Thus, $\mathcal{D}$ consists of C^∞ functions $\varphi(x_1, \ldots x_n)$ that are bounded and stay bounded after every finite succession of partial differentiations and multiplication by x_1, x_2, ... or x_n. Since $\mathcal{D}$ is mapped into itself by the transformation

$$e^{i\alpha\tau\Pi_j}\varphi(x_1, \ldots x_n) = \varphi(x_1 + \tau\delta_{1j}, \ldots x_n + \tau\delta_{nj}),$$

the elements φ of $\mathcal{D}$ cannot all vanish at the same point $x_1, \ldots x_n$.

This being the case, let $g(q) = g(q_1, \ldots q_n)$ be a C^∞ function of the q_j, and let $g(p) = g(p_1, \ldots p_n)$ be the function obtained by substituting p_j for q_j in $g(q)$. We will prove that *there exists one and only one real C^∞ function $F^{(g)}(x_1, \ldots x_n)$ such that*

$$A[g(p)]\varphi = F^{(g)}(\Pi_1, \ldots \Pi_n)\varphi \; ; \quad A[g(q)]\varphi = F^{(g)}(R_1, \ldots R_n)\varphi. \quad (23.2)$$

for all $\varphi \in \mathcal{D}$.

Indeed, the transformations $\gamma_\sigma^{(g(q))}$ commute with $\gamma_\tau^{(q_j)}$ in Γ. Then, by (q_1'),

$$e^{i\tau R_j} e^{i\sigma A[g(q)]} e^{-i\tau R_j} = e^{i\sigma A[g(q)]}$$

for all $j = 1, \ldots n$ and for all real values of σ and τ. By some known theorems (Stone [31] and von Neumann [21]) it follows that there exists a measurable real function $F^{(g)}(x_1, \ldots x_n)$ such that

$$A[g(q)] = F^{(g)}(R_1, \ldots R_n).$$

With (23.1) this gives

$$A[g(q)]\varphi(x_1, \ldots x_n) = F^{(g)}(x_1, \ldots x_n)\varphi(x_1, \ldots x_n).$$

Let $\varphi \in \mathcal{D}$, $\varphi \neq 0$ at the point $x_1, \ldots x_n$. In a neighborhood of this point one has

$$F^{(g)}(x_1, \ldots x_n) = \frac{A[g(q)]\varphi(x_1, \ldots x_n)}{\varphi(x_1, \ldots x_n)}.$$

Also, φ and $A[g(q)]\varphi$ are C^∞ functions belonging to $\mathcal{D}$. It follows that $F^{(g)}(x_1, \ldots x_n)$ is a well-defined C^∞ function.

Let us consider the transformation γ_0 in the group Γ, with equations

$$s' = s - \sum p_j q_j, \quad p'_j = q_j, \quad q'_j = -p_j.$$

This is the transformation $\gamma_{\tau_0}^{(f_0)}$ for $f_0 = \frac{1}{2}\sum(p_j^2 + q_j^2)$ and $\tau_0 = \pi/2$. For all $f \in \mathcal{F}_\Gamma$ and all real τ we have

$$\gamma_0 \gamma_\tau^{(f)} \gamma_0^{-1} = \gamma_\tau^{(\tilde{f})} \qquad (23.3)$$

with

$$\tilde{f}(p_1, \ldots p_n, q_1, \ldots q_n) = f(-q_1, \ldots - q_n, p_1, \ldots p_n) \in \mathcal{F}_\Gamma.$$

If we set

$$U_0 = e^{i\alpha\tau_0 A[f_0]},$$

we deduce from (23.3) and (q'_1) the following relation:

$$U_0 e^{i\alpha\tau A[f]} U_0^{-1} = e^{i\alpha\tau A[\tilde{f}]},$$

hence, for the elements φ of $\mathcal{D}$

$$U_0 A[f] U_0^{-1} \varphi = A[\tilde{f}]\varphi.$$

Applied to the functions $g(p)$, $g(q)$ considered above, this relation gives

$$A[g(p)]\varphi = U_0 A[g(p)] U_0^{-1}\varphi = U_0 F^{(g)}(R_1, \ldots R_n) U_0^{-1}\varphi =$$
$$= F^{(g)}(U_0 R_1 U_0^{-1}, \ldots U_0 R_n U_0^{-1})\varphi = F^{(g)}(\Pi_1, \ldots \Pi_n)\varphi.$$

Thus, the equations (23.2) are proven.

We now prove the following relations[49]

$$F^{(g')}(x_1, \ldots x_n) = \frac{\partial}{\partial x_j} F^{(g)}(x_1, \ldots x_n) \text{ if } g'(q) = \frac{\partial g(q)}{\partial q_j}, \qquad (23.4)$$

$$F^{(\tilde{g})}(x_1, \ldots x_n) = x_j F^{(g)}(x_1, \ldots x_n) \text{ if } \tilde{g}(q) = q_j g(q), p_j g(q) \in \mathcal{F}_\Gamma (23.5)$$

To prove the first expression, note that for $\varphi \in \mathcal{D}$

$$\frac{i}{\alpha}A\left[\frac{\partial}{\partial q_j}g(q)\right]\varphi = \frac{i}{\alpha}A[(g(q),p_j)]\varphi = \{A[g(q)]\Pi_j - \Pi_j A[g(q)]\}\varphi$$
$$= \left\{F^{(g)}(R_1,\ldots R_n)\Pi_j - \Pi_j F^{(g)}(R_1,\ldots R_n)\right\}\varphi,$$

hence, by (23.1)

$$\frac{i}{\alpha}A\left[\frac{\partial}{\partial q_j}g(q)\right]\varphi = \frac{i}{\alpha}\left[\frac{\partial}{\partial x_j}F^{(g)}\right](R_1,\ldots R_n)\varphi.$$

and (23.4) follows.

Now let

$$g(q) = \frac{1}{2}q_j^2.$$

We have $g'(q) = q_j$; $F^{(g')}(x_1,\ldots x_n)$ is determined by

$$F^{(g')}(R_1,\ldots R_n)\varphi = A[q_j]\varphi = R_j\varphi, \; (\varphi \in \mathcal{D})$$

which gives $F^{(g')}(x_1,\ldots x_n) = x_j$. With the help of (23.4) we find

$$F^{(g)}(x_1,\ldots x_n) = \frac{x_j^2}{2} + a$$

and

$$A\left[\frac{1}{2}p_j^2\right]\varphi = \left(\frac{1}{2}\Pi_j^2 + aI\right)\varphi, \quad A\left[\frac{1}{2}q_j^2\right]\varphi = \left(\frac{1}{2}R_j^2 + aI\right)\varphi, \quad (23.6)$$

where a is a real number and I is the identity transformation of the space $\mathcal{H}_n$.

Again, set

$$g'(q) = \frac{\partial}{\partial q_j}g(q),$$

and $p_j g'(q) \in \mathcal{F}_\Gamma$; we can write, using (23.6), ($\varphi \in \mathcal{D}$),

$$A[p_j g'(q)]\varphi = A\left[\left(g(q),\frac{1}{2}p_j^2\right)\right]\varphi = \frac{\alpha}{2i}\{A[g(q)]\Pi_j^2 - \Pi_j^2 A[g(q)]\}\varphi$$
$$= \frac{\alpha}{2i}\left[\{A[g(q)]\Pi_j - \Pi_j A[g(q)]\}\Pi_j + \Pi_j\{A[g(q)]\Pi_j - \Pi_j A[g(q)]\}\right]\varphi$$
$$= \frac{1}{2}\{A[(g(q),p_j)]\Pi_j + \Pi_j A[(g(q),p_j)]\}\varphi,$$

hence

$$A[p_j g'(q)]\varphi = \frac{1}{2}\{A[g'(q)]\Pi_j + \Pi_j A[g'(q)]\}\varphi. \quad (23.7)$$

Next,
$$A[q_j g'(q)]\varphi = A\left[\left(\frac{1}{2}q_j^2, p_j g'(q)\right)\right]\varphi =$$
$$= \frac{\alpha}{2i}\left\{R_j^2 A[p_j g'(q)] - A[p_j g'(q)]R_j^2\right\}\varphi .$$

We now substitute (23.7), note that $A[g'(q)]$ commutes with R_j, and use the relation
$$\frac{\alpha}{2i}\left\{R_j^2 \Pi_j - \Pi_j R_j^2\right\}\varphi = R_j \varphi .$$

Then, it follows that
$$A[q_j g'(q)]\varphi = R_j A[g'(q)]\varphi .$$

In this relation, we can replace $g'(q)$ by any C^∞ function in $q_1, \ldots q_n$ satisfying $p_j g(q) \in \mathcal{F}_\Gamma$. Using (23.2), we see that the above relation is equivalent to (23.5).

Using (23.4) and (23.5), we now look for a function $F_0(x_1, \ldots x_n) = F^{(g_0)}(x_1, \ldots x_n)$ with $g_0(q) = e^{-q_j^2/2}$. We have
$$\frac{\partial g_0}{\partial q_j} + q_j g_0 = 0, \quad \frac{\partial g_0}{\partial q_k} = 0, \quad (k \neq j) .$$

From (23.4) and (23.5) it follows that F_0 is the solution of
$$\frac{\partial F_0}{\partial x_j} + x_j F_0 = 0, \quad \frac{\partial F_0}{\partial x_k} = 0, \quad (k \neq j) .$$

Thus
$$F_0 = c e^{-x_j^2/2} ,$$
where c is a real constant which we will not attempt to fix. Further, set $g_1(q) = \phi(q_j)$ with $\phi(x) = \int_{-\infty}^{x} e^{-u^2/2}\, du$. We have, by (23.4),
$$F^{(g_1)}(x_1, \ldots x_n) = c\phi(x_j) + d ,$$
d being a real constant. Now, apply (23.7) with
$$g(q) = g_0(q) = e^{-q_j^2/2} .$$

By (23.2) and the value we found for F_0, this gives for $\varphi \in \mathcal{D}$
$$A[p_j g_0(q)]\varphi = \frac{c}{2}\left\{\Pi_j g_0(R) + g_0(R)\Pi_j\right\}\varphi .$$

If
$$\Psi(x) = \int_0^x e^{u^2/2} du ,$$
we have
$$\Psi(q_j) \in \mathcal{F}_\Gamma$$
and
$$\Pi_j \varphi = A[p_j]\varphi = -A[(p_j g_0(q), \Psi(q_j))]\varphi =$$
$$-\frac{\alpha}{i} \{A[p_j g_0(q)]A[\Psi(q_j)] - A[\Psi(q_j)]A[p_j g_0(q)]\} \varphi .$$

We cannot have $\Pi_j \varphi = 0$. It follows from the expression obtained for $A[p_j g_0(q)]$ that the constant c cannot be zero. By our hypotheses, we arrive at the following conclusion. *If the function $\phi(x)$ is defined by*
$$\phi(x) = \int_{-\infty}^x e^{-u^2/2} du ,$$
we have for $j = 1, \ldots n$ and $\varphi \in \mathcal{D}$
$$A[\phi(p_j)]\varphi = (c\phi(\Pi_j) + c'I)\varphi , \quad A[\phi(q_j)]\varphi = (c\phi(R_j) + c'I)\varphi , \qquad (23.8)$$
where $c \neq 0$ and c' are real constants.

Using (23.8) we now have ($\varphi \in \mathcal{D}$)
$$A[e^{-\frac{1}{2}(p_j^2+q_j^2)}]\varphi = A[(\phi(q_j), \phi(p_j))]\varphi \qquad (23.9)$$
$$= \frac{\alpha}{i} \{A[\phi(q_j)]A[\phi(p_j)] - A[\phi(p_j)]A[\phi(q_j)]\} \varphi$$
$$= \frac{\alpha c^2}{i} \{\phi(R_j)\phi(\Pi_j) - \phi(\Pi_j)\phi(R_j)\} \varphi .$$

In the group Γ, the transformations $\gamma_\sigma^{(f)}$ with $f = e^{-\frac{1}{2}(p_j^2+q_j^2)}$ commute with the transformations $\gamma_\tau^{(g)}$ with $g = \frac{1}{2}(p_j^2 + q_j^2)$. *Thus, the operator (23.9) must commute with the transformations $e^{i\alpha\tau A[\frac{1}{2}(p_j^2+q_j^2)]}$.* The last part of the proof will show that this cannot be the case. To this end, we will express the operator (23.9) using an integral of operators. Let $F(u)$ be an integrable function on $-\infty < u < +\infty$:
$$\int_{-\infty}^{+\infty} |F(u)| du < +\infty ;$$
the Fourier transform
$$G(u') = \frac{1}{\sqrt{2\pi}} \int_{-\infty}^{+\infty} e^{iuu'} F(u) du$$

is a continuous bounded function. We have the following equalities between bounded operators

$$G(\Pi_j) = \frac{1}{\sqrt{2\pi}} \int_{-\infty}^{+\infty} F(u) e^{iu\Pi_j} du \,; \quad G(R_j) = \frac{1}{\sqrt{2\pi}} \int_{-\infty}^{+\infty} F(v) e^{ivR_j} dv \,,$$

that one easily verifies from (23.1). For the definition of integrals of operators, one can consult von Neumann [21], for example. We find

$$G(R_j)G(\Pi_j) - G(\Pi_j)G(R_j) =$$
$$\frac{1}{2\pi} \int_{-\infty}^{+\infty} du \int_{-\infty}^{+\infty} dv \, F(u)F(v) \left(e^{ivR_j} e^{iu\Pi_j} - e^{iu\Pi_j} e^{ivR_j} \right) \,.$$

Furthermore, by (23.1)

$$e^{i(u\Pi_j + vR_j)} = e^{\frac{i}{2\alpha}uv} e^{ivR_j} e^{iu\Pi_j} = e^{-\frac{i}{2\alpha}uv} e^{iu\Pi_j} e^{ivR_j}$$

If we set $s(x) = x^{-1} \sin x$, we find

$$\frac{\alpha}{i} \{G(R_j)G(\Pi_j) - G(\Pi_j)G(R_j)\} = \qquad (23.10)$$
$$-\frac{1}{2\pi} \int_{-\infty}^{+\infty} du \int_{-\infty}^{+\infty} dv \, uF(u) \cdot vF(v) \cdot s\left(\frac{uv}{2\alpha}\right) e^{i(u\Pi_j + vR_j)} \,.$$

Apply this relation to the function $F(u) = -i(u - i\epsilon)^{-1} e^{-(u-i\epsilon)^2/2}$, where ϵ is a real positive number. Then, the function $G(u')$ is $e^{-u'\epsilon} \phi(u')$ where $\phi(u') = \int_{-\infty}^{u'} e^{-x^2/2} dx$. Now, let ϵ tend to zero. The operators on both sides of (23.10) strongly converge to limits that give rise to the following relation

$$\frac{\alpha}{i} \{\phi(R_j)\phi(\Pi_j) - \phi(\Pi_j)\phi(R_j)\} =$$
$$\frac{1}{2\pi} \int_{-\infty}^{+\infty} du \int_{-\infty}^{+\infty} dv \, e^{-\frac{1}{2}(u^2+v^2)} s\left(\frac{uv}{2\alpha}\right) e^{i(u\Pi_j + vR_j)} \,.$$

Relation (23.9) now gives, for $\varphi \in \mathcal{D}$

$$A[e^{-\frac{1}{2}(p_j^2 + q_j^2)}]\varphi = \qquad (23.11)$$
$$\frac{c^2}{2\pi} \left\{ \int_{-\infty}^{+\infty} du \int_{-\infty}^{+\infty} dv \, e^{-\frac{1}{2}(u^2+v^2)} s\left(\frac{uv}{2\alpha}\right) e^{i(u\Pi_j + vR_j)} \right\} \varphi \,.$$

As we noted above, for $f = e^{-\frac{1}{2}(p_j^2 + q_j^2)}$ and all real τ, we have by condition (q_1')

$$e^{i\alpha\tau A[f]} A[e^{-\frac{1}{2}(p_j^2 + q_j^2)}] e^{-i\alpha\tau A[f]} = A[e^{-\frac{1}{2}(p_j^2 + q_j^2)}] \,.$$

Now, with this choice of f, an easy calculation gives
$$e^{i\alpha\tau A[f]}e^{i(u\Pi_j+vR_j)}e^{-i\alpha\tau A[f]} = e^{i(u'\Pi_j+v'R_j)}$$
with
$$u' = u\cos\tau + v\sin\tau, \quad v' = -u\sin\tau + v\cos\tau.$$
Hence
$$A[e^{-\frac{1}{2}(p_j^2+q_j^2)}]\varphi =$$
$$\frac{c^2}{2\pi}\left\{\int_{-\infty}^{+\infty} du \int_{-\infty}^{+\infty} dv\, e^{-\frac{1}{2}(u^2+v^2)}\, s\left(\frac{uv}{2\alpha}\right)e^{i(u'\Pi_j+v'R_j)}\right\}\varphi.$$

Make a change of variables in the double integral, from u, v, to u', v', yielding
$$A[e^{-\frac{1}{2}(p_j^2+q_j^2)}]\varphi =$$
$$\frac{c^2}{2\pi}\left\{\int_{-\infty}^{+\infty} du' \int_{-\infty}^{+\infty} dv'\, e^{-\frac{1}{2}(u'^2+v'^2)}\, s\left(\frac{uv}{2\alpha}\right)e^{i(u'\Pi_j+v'R_j)}\right\}\varphi,$$
where uv should be replaced by $(u'\cos\tau - v'\sin\tau)(u'\sin\tau + v'\cos\tau)$. Comparing with (23.11), let us now set
$$r(u,v) = \tag{23.12}$$
$$e^{-\frac{1}{2}(u^2+v^2)}\left\{s\left[\frac{1}{2\alpha}(u\cos\tau - v\sin\tau)(u\sin\tau + v\cos\tau)\right] - s\left(\frac{uv}{2\alpha}\right)\right\}.$$

Since $c \neq 0$, it follows
$$\left\{\int_{-\infty}^{+\infty} du \int_{-\infty}^{+\infty} dv\, r(u,v)e^{i(u\Pi_j+vR_j)}\right\}\varphi = 0,$$
that is, for $\varphi \in \mathcal{D}$, $\psi \in \mathcal{H}_n$
$$\int_{-\infty}^{+\infty} du \int_{-\infty}^{+\infty} dv\, r(u,v)(\psi, e^{i(u\Pi_j+vR_j)}\varphi) = 0.$$
Replace φ and ψ by
$$e^{i(u'\Pi_j+v'R_j)}\varphi \quad \text{and} \quad e^{i(u'\Pi_j+v'R_j)}\psi,$$
respectively, with u' and v' real. The above equality becomes
$$\int_{-\infty}^{+\infty} du \int_{-\infty}^{+\infty} dv\, r(u,v)(\psi, e^{i(u\Pi_j+vR_j)}\varphi)e^{\frac{i}{\alpha}(uv'-u'v)} = 0.$$
Since it must be true for all values of u', v', we must have
$$r(u,v)(\psi, e^{i(u\Pi_j+vR_j)}\varphi) = 0.$$

for all real values of u, v. Taking $\psi = e^{i(u\Pi_j + v R_j)}\varphi$, we conclude that $r(u,v) = 0$ for all values u and v. By (23.12), this is manifestly not true for $\tau \neq 0$ and $\tau \neq \pi$. The hypothesis of existence of a map $f \to A[f]$ satisfying the conditions (q_1') and (q_2') thus leads to a contradiction. The theorem is proven.

Using theorems by Stone and von Neumann reviewed in section 21, the theorem we just proved gives the following corollary.

COROLLARY: *There is no map $f \to A[f]$ of $\mathcal{F}_\Gamma$ in the set of hypermaximally symmetric operators of a separable Hilbert space $\mathcal{H}$ such that the condition (q_1') is satisfied and the operators $e^{i\alpha\tau A[f]}$ (τ real, $f \in \mathcal{F}_\Gamma$) all belong to the ring of bounded operators generated by the operators $e^{i\alpha\tau A[p_j]}$, $e^{i\alpha\tau A[q_j]}$, (τ real, $j = 1\ldots n$).*

The corollary is also proven *ad absurdum*. We assume there exist a map $f \to A[f]$ with the stated properties. By the theorems of Stone and von Neumann, $\mathcal{H}$ contains a subspace $\mathcal{H}'$ invariant under $e^{i\tau A[p_j]}$, $e^{i\tau A[q_j]}$, which can be identified with the space $\mathcal{H}_n$ in the same way that $e^{i\tau A[p_j]}$, $e^{i\tau A[q_j]}$ reduce to $e^{i\tau \Pi_j}$, $e^{i\tau R_j}$. The space $\mathcal{H}' \equiv \mathcal{H}_n$ is invariant under all transformations $e^{i\tau A[f]}$, ($f \in \mathcal{F}_\Gamma$), and the restrictions of $A[f]$ to the orthogonal projection of the domain $\mathcal{D}$ in $\mathcal{H}'$ satisfy conditions (q_1'), (q_2'). This brings us to the above corollary.

We note that for the map $f \to A[f] \equiv \frac{1}{\alpha} H^{(\alpha)}[f]$, which as we know satisfies condition (q_1'), the operators $e^{i\alpha\tau A[f]}$ belong to the ring generated by the union of the ring $\mathcal{A}$ that generates the operators $e^{i\tau A[p_j]}$, $e^{i\tau A[q_j]}$, and a ring $\tilde{\mathcal{A}}$ which is isomorphic to $\mathcal{A}$ and identical to its commutator (see section 22). In section 22, we reviewed the bijective correspondence between classical and quantum quantities proposed by Weyl [34]. An explicit calculation shows that for f_1, f_2, $(f_1, f_2) \in \mathcal{F}_\Gamma$ we do not always have

$$A[f_1]A[f_2] - A[f_2]A[f_1] = \frac{i}{\alpha} A[(f_1, f_2)], \quad (\alpha = 2\pi/h). \tag{23.13}$$

For details, we refer to previously mentioned work of Groenewold [13] (p. 46-49).[t50] The differences that appear between the two sides of (23.13) are always second order in $h/2\pi = \alpha^{-1}$. This is in accord with the fact that asymptotically, the formalism of Quantum Mechanics should converge to that of Classical Mechanics when $h \to 0$.

24 Bijective Correspondence Between Quadratic Quantities

The situation described in the theorem of the preceding paragraph is completely changed when, instead of considering all infinitesimal transformations of the group Γ, one restricts attention to those of the subgroup L. We know that the infinitesimal transformations $X[f]$ in L correspond to elements f of the family $\mathcal{F}_L$ of real polynomials of degree 0, 1 and 2 in the p_j, q_j. The space $\mathcal{H}_n$ and the operators Π_j, R_j being defined as in the previous section, we state

THEOREM I: *For every real α, there exists a map $f \to A[f]$ of $\mathcal{F}_L$ in the set of hypermaximally symmetric operators on the space $\mathcal{H}_n$ with the following properties (q_1'') and (q_2'').*[t51]

(q_1'') *There exists a domain Δ_n, dense in $\mathcal{H}_n$, with the following properties:*

- *it is common to all domains of definition of the operators $A[f]$, ($f \in \mathcal{F}_L$)*

- *every $A[f]$ and every $e^{i\alpha\tau A[f]}$ ($f \in \mathcal{F}_L$, τ real) map Δ_n into itself*

- *the restriction of every $A[f]$ ($f \in \mathcal{F}_L$) to Δ_n is hypermaximally symmetric*

- *for $\chi \in \Delta_n$, f_1 and $f_2 \in \mathcal{F}_L$, a_1, a_2 real, we have*

$$\begin{cases} \{a_1 A[f_1] + a_2 A[f_2]\}\chi = A[a_1 f_1 + a_2 f_2]\chi \, ; \\ \{A[f_1]A[f_2] - A[f_2]A[f_1]\}\chi = \frac{i}{\alpha} A[(f_1, f_2)]\chi \end{cases} \quad (24.1)$$

- *the map $\gamma_\tau^{(f)} \to e^{i\alpha\tau A[f]}$ ($f \in \mathcal{F}_L$, τ real) satisfies*

$$\gamma_\sigma^{(g')} = \gamma_\tau^{(f)}\gamma_\sigma^{(g)}\gamma_{-\tau}^{(f)} \to e^{i\alpha\sigma A[g']} = e^{i\alpha\tau A[f]}e^{i\alpha\sigma A[g]}e^{-i\alpha\tau A[f]} \quad (24.2)$$

for all f, g, $g' \in \mathcal{F}_L$ and σ, τ real.

(q_2'') *For $j = 1, \ldots n$, the operators $A[p_j]$, $A[q_j]$ are identical to the operators Π_j, R_j, respectively.*

As in the theorem of section 23, the last part of condition (q_1'') can be replaced by the condition that the maps $f \to A[f]$, $\gamma_\tau^{(f)} \to e^{i\alpha\tau A[f]}$ furnish a representation in $\mathcal{H}_n$ of the adjoint group[t52] of L.

In order to prove the theorem using the operators $H^{(\alpha)}[f]$ from chapter IV, we first need a lemma. $\mathcal{H}_n$ denotes the Hilbert space of functions $\chi(x_1, \ldots x_n)$, measurable and square integrable on the n-dimensional Euclidean space (x), with volume element $dx_1, \ldots dx_n$. On the other hand, $\mathcal{H}_{2n}$ denotes the Hilbert space of functions $\varphi(p_1, \ldots p_n, q_1, \ldots q_n)$, measurable and square integrable on the $2n$-dimensional Euclidean space (p, q), with volume element $dp_1, \ldots dp_n \, dq_1, \ldots dq_n$. Our lemma is the following:

LEMMA: *Let $\mathcal{D}_p$, $\mathcal{D}_q$ be two dense domains in $\mathcal{H}_n$, and $\mathcal{D}$ a dense domain in $\mathcal{H}_{2n}$ such that if $\varphi(p_1, \ldots p_n, q_1, \ldots q_n) \in \mathcal{D}$, we have $\varphi(x_1, \ldots x_n, q_1, \ldots q_n) \in \mathcal{D}_p$ for almost all points $q_1, \ldots q_n$ and $\varphi(p_1, \ldots p_n, x_1, \ldots x_n) \in \mathcal{D}_q$ for almost all points $p_1, \ldots p_n$. Let A_p, A_q be two* linear symmetric operators in $\mathcal{H}_n$, defined on $\mathcal{D}_p$, $\mathcal{D}_q$ respectively, such that $A = A_p + A_q$ (where A_p acts on the variables $p_1, \ldots p_n$ and A_q on $q_1, \ldots q_n$) is a symmetric operator with a domain of definition that contains $\mathcal{D}$. If the restriction of A to $\mathcal{D}$ is essentially hypermaximally symmetric, the operators A_p, A_q, defined on $\mathcal{D}_p$, $\mathcal{D}_q$ respectively, have the same property.

To prove this, we first note that if a_1 and a_2 are two real numbers of nonzero sum, one of the sets $(A_p + ia_1)\mathcal{D}_p$ or $(A_q + ia_2)\mathcal{D}_q$ is dense in $\mathcal{H}_n$. If this were not true, we could find an element $\chi_p(x_1, \ldots x_n)$ orthogonal to $(A_p + ia_1)\mathcal{D}_p$ and an element $\chi_q(x_1, \ldots x_n)$ orthogonal to $(A_q + ia_2)\mathcal{D}_q$. Then, for all $\varphi \in \mathcal{D}$ and $a = a_1 + a_2$, the element $\psi = \chi_p(p_1, \ldots p_n) \cdot \chi_q(q_1, \ldots q_n)$ satisfies

$$(\psi, (A + ia)\varphi) = (\psi, (A_p + ia_1)\varphi) + (\psi, (A_q + ia_2)\varphi) = 0 ,$$

the two terms after the first equal sign vanishing separately. Thus, ψ is orthogonal to $(A + ia)\mathcal{D}$ for a real and nonzero, which is impossible (see section 13, lemma I).

From this remark results that at least one of the operators A_p, A_q is essentially hypermaximal. Indeed, suppose A_p does not have this property. One knows, then, that one of the domains $(A_p \pm i)\mathcal{D}_p$ is not dense in $\mathcal{H}_n$ (see Stone[30], p. 339). Thus, the two domains $(A_q \pm 2i)\mathcal{D}_q$ are dense in $\mathcal{H}_n$, which means A_q is essentially hypermaximal.

From this point on, we can assume that A_q is essentially hypermaximal. Without loss of generality, we can assume zero to be an eigenvalue, or an accumulation point of the spectrum, of the hypermaximal extension A_q^* of A_q. Then, we can find an element χ_q' of $\mathcal{H}_n$ such that $A_q^* \chi_q'$ exists and satisfies $\|A_q^* \chi_q'\| < \frac{1}{2}\|\chi_q'\|$. If A_p is not essentially hypermaximal, $\mathcal{H}_n$ contains an element χ_p' orthogonal to one of its domains $(A_p \pm i)\mathcal{D}_p$, say $(A_p + i)\mathcal{D}_p$. Let $\psi' = \chi_p'(p_1, \ldots p_n) \cdot \chi_q(q_1, \ldots q_n)$. We have for all $\varphi \in \mathcal{D}$

$$(\psi', (A + ia)\varphi) = (\psi', (A_p + i)\varphi) + (\psi', A_q \varphi) = (\psi', A_q \varphi) = (\tilde{\psi}, \varphi)$$

with

$$\tilde{\psi} = \chi_p'(p_1, \ldots p_n) \cdot A_q^* \chi_q'(x_1, \ldots x_n) .$$

Hence, with the norm of each element being taken in their respective spaces,

$$|(\psi', (A + ia)\varphi)| \leq \|\varphi\| \cdot \|\tilde{\psi}\| = \quad (24.3)$$

$$\|\varphi\| \cdot \|\chi_p'\| \cdot \|A_q^* \chi_q'\| < \frac{1}{2} \|\varphi\| \cdot \|\chi_p'\| \cdot \|\chi_q'\| .$$

Since $(A + i)\mathcal{D}$ is dense in $\mathcal{H}_{2n}$, we can find $\varphi \in \mathcal{D}$ such that

$$|(\psi', (A + i)\varphi)| > \frac{1}{2} \|(A + i)\varphi\| \cdot \|\psi'\| .$$

Now,
$$\|\psi'\| = \|\chi'_p\| \cdot \|\chi'_q\| \quad \text{and} \quad \|(A+i)\varphi\|^2 = \|A\varphi\|^2 + \|\varphi\|^2 \ .$$
Hence
$$|(\psi', (A+i)\varphi)| > \frac{1}{2}\|\varphi\| \cdot \|\chi'_p\| \cdot \|\chi'_q\| \ . \tag{24.4}$$

The inequalities (24.3) and (24.4) being contradictory, A_p must also be essentially hypermaximal. The lemma is established. We note that it constitutes a special case of a more general proposition concerning direct products of two Hilbert spaces, which is proven in an analogous manner.

To apply the lemma to the operators $H^{(\alpha)}[f]$, we recall the results and notation of section 16. Take an element of $\mathcal{F}_L$
$$f = \sum a_{jk} p_j p_k + \sum b_{jk} q_j q_k + 2\sum c_{jk} p_j q_k + \sum a_j p_j + \sum b_j q_j + d \ ,$$
with
$$a_{jk} = a_{kj} \ , \quad b_{jk} = b_{kj} \ .$$

The formula (16.9) gives
$$\frac{1}{\alpha} H^{(\alpha)}[f] = F^{(f)}(P, Q) + F_1^{(f)}(P', Q')$$
with
$$F^{(f)}(P, Q) = \sum a_{jk} P_j P_k + \sum b_{jk} Q_j Q_k + \sum c_{jk}(P_j Q_k + Q_k P_j)$$
$$+ \sum a_j P_j + \sum b_j Q_j + d \ ,$$
$$F_1^{(f)}(P', Q') = -\sum a_{jk} P'_j P'_k - \sum b_{jk} Q'_j Q'_k + \sum c_{jk}(P'_j Q'_k + Q'_k P'_j) \ .$$

This relation is valid on the domain Δ_{2n} of C^∞ functions $\varphi(p_1, \ldots p_n, q_1, \ldots q_n)$ on the space (p, q) that are bounded and stay bounded after every finite succession of partial differentiations and multiplication by $p_1, \ldots p_n, q_1, \ldots q_n$. The restriction of the operator $\frac{1}{\alpha} H^{(\alpha)}[f]$ to this domain is essentially hypermaximally symmetric. Since the transformation W defined in (16.1) maps Δ_{2n} into itself, the restriction of the operator $\frac{1}{\alpha} W^{-1} H^{(\alpha)}[f] W$ to Δ_{2n} is also essentially hypermaximally symmetric. By the previous relation, and with the same definition of the functions of operators $F^{(f)}$, $F_1^{(f)}$, we have
$$\frac{1}{\alpha} W^{-1} H^{(\alpha)}[f] W = F^{(f)}(A, B) + F_1^{(f)}(A', B') \ . \tag{24.5}$$

The operators A_j, B_j, A'_j, B'_j are defined in (16.2), (16.3); they are related to P_j, Q_j, P'_j, Q'_j by (16.4). Whereas the A_j, B_j, act on the variables $q_1, \ldots q_n$, the A'_j, B'_j, act on the variables $p_1, \ldots p_n$. Thus, we can apply the lemma.

Let us now turn to the Hilbert space $\mathcal{H}_n$ and the operators Π_j, R_j defined in (23.1).[t53] Let Δ_n be the domain of $\mathcal{H}_n$ of C^∞ functions $\chi(x_1, \ldots x_n)$, that are bounded and stay bounded after every finite succession of partial differentiations and multiplication by $x_1, \ldots x_n$. The domain Δ_n is contained in the domains of definition of the symmetric operators

$$F^{(f)}(\Pi, R) = \quad (24.6)$$
$$\sum a_{jk} \Pi_j \Pi_k + \sum b_{jk} R_j R_k + \sum c_{jk}(\Pi_j R_k + R_k \Pi_j) \sum a_j \Pi_j + \sum b_j R_j + d,$$
$$F_1^{(f)}(\Pi, R) = \quad (24.7)$$
$$-\sum a_{jk} \Pi_j \Pi_k - \sum b_{jk} R_j R_k + \sum c_{jk}(\Pi_j R_k + R_k \Pi_j).$$

By the lemma, *the restrictions to Δ_n of the operators $F^{(f)}(\Pi, R)$, $F_1^{(f)}(\Pi, R)$ are essentially hypermaximal*. One can also state this conclusion in the following form. *Every polynomial of second degree in the operators $\Pi_1, \ldots \Pi_n$, $R_1, \ldots R_n$, considered as operator defined on the domain Δ_n, is essentially hypermaximally symmetric, provided it is formally identical to its adjoint* (the latter being obtained by exchanging the order of factors Π_j, R_j in each term). This proposition can be proven without recourse to our knowledge of the operators $H^{(\alpha)}[f]$, but instead applying to the subgroup T of L a suitably completed theorem of Segal ([28], theorem 3.1). We can show by counterexample that the proposition does not extend to polynomials of arbitrary degree.[k] [t54]

We now verify that the map

$$f \to A[f] = F^{(f)}(\Pi, R), \quad (f \in \mathcal{F}_L)$$

satisfies the theorem above. By (24.6) it satisfies the property (q''_2). The relations (24.1) are easily checked by calculation. Thus, it remains to show that the transformations $e^{i\alpha\tau A[f]}$, $(f \in \mathcal{F}_L, \tau$ real) map Δ_n into itself, and to verify property (24.2).

To prove the first point, consider two elements $\chi(x_1, \ldots x_n)$, $\chi'(x_1, \ldots x_n)$ of Δ_n. The function

$$\psi(p_1, \ldots p_n, q_1, \ldots q_n) = \chi(q_1, \ldots q_n)\chi'(p_1, \ldots p_n) \quad (24.8)$$

belongs to the domain Δ_{2n}. We know that it is invariant under the transformation W and under transformations in the representation $\mathcal{R}^{(\alpha)}$ of L. Therefore,

---

[k]Private communication to the author by J. VON NEUMANN.

for $f \in \mathcal{F}_L$ and τ real, it will also contain the element[55]

$$W^{-1}e^{i\tau H^{(\alpha)}[f]}W\psi = \qquad (24.9)$$
$$e^{i\tau W^{-1}H^{(\alpha)}[f]W}\psi = e^{i\alpha\tau F^{(f)}(A,B)}e^{i\alpha\tau F_1^{(f)}(A',B')}\psi$$

which is none other than the function

$$e^{i\alpha\tau F^{(f)}(A,B)}\chi(q_1,\ldots q_n) \cdot e^{i\alpha\tau F_1^{(f)}(A',B')}\chi'(p_1,\ldots p_n) .$$

Let us integrate this function over a compact set $\mathcal{E}$ of the space $p_1,\ldots p_n$ such that

$$\int_{\mathcal{E}} e^{i\alpha\tau F_1^{(f)}(A',B')}\chi'(p_1,\ldots p_n)\, dp_1\ldots dp_n = m \neq 0 .$$

One obtains the function

$$m e^{i\alpha\tau F^{(f)}(A,B)}\chi(q_1,\ldots q_n)$$

which will be C^∞ in $q_1,\ldots q_n$, and it will be bounded and stay bounded after every finite succession of partial differentiations and multiplication by $q_1,\ldots q_n$. Now let us replace the variables q_j by x_j, and the operators A_j, B_j by Π_j, R_j, respectively. Since $A[f] = F^{(f)}(\Pi, R)$, we have shown that $e^{i\alpha\tau A[f]}\chi \in \Delta_n$ if $\chi \in \Delta_n$, $f \in \mathcal{F}_L$ and τ real.

We now turn to property (24.2). Let $f, g, g' \in \mathcal{F}_L$, σ and τ real be such that

$$\gamma_\sigma^{(g')} = \gamma_\tau^{(f)}\gamma_\sigma^{(g)}\gamma_{-\tau}^{(f)} . \qquad (24.10)$$

If we denote points in the space (p,q) by ω, then g' is related to g by $g'(\omega) = g(\gamma_{-\tau}^{(f)}\omega)$. In the representation $\mathcal{R}^{(\alpha)}$, relation (24.10) becomes

$$e^{i\sigma H^{(\alpha)}[g']} = e^{i\tau H^{(\alpha)}[f]}e^{i\sigma H^{(\alpha)}[g]}e^{-i\tau H^{(\alpha)}[f]}$$

and gives for every element φ of Δ_{2n}

$$H^{(\alpha)}[g']\varphi = e^{i\tau H^{(\alpha)}[f]}H^{(\alpha)}[g]e^{-i\tau H^{(\alpha)}[f]}\varphi .$$

Let us transform this equation by W and apply (24.5) and (24.9). It follows for all $\varphi \in \Delta_{2n}$

$$F^{(g')}(A,B)\varphi + F_1^{(g')}(A',B')\varphi =$$
$$e^{i\alpha\tau F^{(f)}(A,B)}\,F^{(g)}(A,B)\,e^{-i\alpha\tau F^{(f)}(A,B)}\varphi +$$
$$e^{i\alpha\tau F_1^{(f)}(A',B')}\,F_1^{(g)}(A',B')\,e^{-i\alpha\tau F_1^{(f)}(A',B')}\varphi .$$

If g is of first order in the variables p_j, q_j, the same holds for g' and by (24.7) $F_1^{(g)} = F_1^{(g')} = 0$. We thus have, in this case,

$$F^{(g')}(A,B)\varphi = e^{i\alpha\tau F^{(f)}(A,B)}\, F^{(g)}(A,B)\, e^{-i\alpha\tau F^{(f)}(A,B)}\varphi\ .$$

Taking for φ an element of the form (24.8), one finds that the previous relation remains satisfied when one replaces A_j, B_j by Π_j, R_j, respectively, and φ by an arbitrary element of Δ_n. Hence, for g of first order, the equation

$$A[g'] = e^{i\alpha\tau A[f]} A[g] e^{-i\alpha\tau A[f]} \tag{24.11}$$

is valid on the domain Δ_n of the space $\mathcal{H}_n$.

Now let g be any element of $\mathcal{F}_L$; again set $g'(\omega) = g(\gamma_{-\tau}^{(f)}\omega)$. Let p_j, q_j be the coordinates of a point of the space (p, q) and p'_j, q'_j the coordinates of its transform by $\gamma_{-\tau}^{(f)}$. We have $g'(p_1, \ldots p_n, q_1, \ldots q_n) = g(p'_1, \ldots p'_n, q'_1, \ldots q'_n)$. Thus, by (24.6) we will have $A[g'] = F^{(g')}(\Pi_j, R_j) = F^{(g)}(\Pi'_j, R'_j)$ where the Π'_j, R'_j have the same expression (of first degree) in the Π_j, R_j that the p'_j, q'_j have in the p_j, q_j. By (24.11) for $g = p_j$ or q_j,

$$\Pi'_j = e^{i\alpha\tau A[f]}\Pi_j e^{-i\alpha\tau A[f]} \qquad R'_j = e^{i\alpha\tau A[f]}R_j e^{-i\alpha\tau A[f]}$$

and it follows

$$A[g'] = F^{(g)}(\Pi'_j, R'_j) = e^{i\alpha\tau A[f]} F^{(g)}(\Pi_j, R_j) e^{-i\alpha\tau A[f]}\ ,$$

hence again relation (24.11) but this time valid for any $g \in \mathcal{F}_L$. The equalities between operators written above, (24.11) in particular, are valid on the domain Δ_n. The restrictions of the operators appearing in this relation to Δ_n are essentially hypermaximally symmetric, so the relation remains valid for hypermaximal extensions of these operators. By (24.11) one can then conclude

$$e^{i\alpha\sigma A[g']} = e^{i\alpha\tau A[f]} e^{i\alpha\sigma A[g]} e^{-i\alpha\tau A[f]}\ .$$

The validity of (24.2) and of our theorem is thus proven.

The theorem I was an existence theorem. We now prove a uniqueness theorem.

THEOREM II: *For every real α, there exists only one map $f \to A[f]$ from $\mathcal{F}_L$ into the set of hypermaximally symmetric operators on $\mathcal{H}_n$ that satisfy conditions (q''_1) and (q''_2) of theorem I.*

Indeed, assume there are two such maps $f \to A[f]$, $f \to A'[f]$. The condition (q''_2) gives $A[p_j] = A'[p_j]$, $A[q_j] = A'[q_j]$. With relations (24.1) this gives $A[g] = A'[g]$ for all g of degree 0 or 1 in the p_j, q_j. These equations are also valid for hypermaximal extensions of the operators they relate. Next, using (24.2) with $g = p_j$, q_j, we find for all $f \in \mathcal{F}_L$, σ and τ real

$$e^{i\alpha\tau A[f]} e^{i\alpha\sigma \Pi_j} e^{-i\alpha\tau A[f]} = e^{i\alpha\tau A'[f]} e^{i\alpha\sigma \Pi_j} e^{-i\alpha\tau A'[f]}.$$

and an analogous relation for R_j. Hence,

$$e^{i\alpha\tau A'[f]} = c(\tau, f) e^{i\alpha\tau A[f]}$$

where $c(\tau, f)$ is a complex number that, by strong continuity of the above operators, is continuous in τ. Further, it satisfies

$$c(\sigma + \tau, f) = c(\tau, f) \cdot c(\sigma, f) \; ; \quad \text{hence} \quad c(\tau, f) = e^{i\alpha\tau\mu[f]}$$

and

$$A'[f] = A[f] + \mu[f]$$

where $\mu[f]$ is a real function of f; by the first relation (24.1) it is linear. The second relation gives $\mu[(f_1, f_2)] = 0$ for any f_1 and f_2 in $\mathcal{F}_L$. Now consider the group L. The infinitesimal transformations $X[(f_1, f_2)]$ generate its derived group L' which is an invariant subgroup. All elements $g \in \mathcal{F}_L$ of degree 0 or 1 can easily be brought to the form (f_1, f_2). Thus, L' contains the invariant subgroup T, generated by the $X[g]$ for g of degree 0 or 1. Since there exist expressions (f_1, f_2) of degree 2, L'/T is a continuous invariant subgroup of the symplectic group L'/T. Now, the symplectic group does not have any nontrivial invariant continuous subgroup (see Dieudonné[6] p. 12). Thus $L = L'$ and $\mu[f]$ is zero for all $f \in \mathcal{F}_L$. Theorem II is proven.

In stating theorems I and II, one can replace the condition (q_2'') by the following: the operators $e^{i\alpha\tau A[f]}$, $(f \in \mathcal{F}_L, \tau \text{ real})$ belong to the ring generated by the operators $e^{i\alpha\tau A[p_j]}$, $e^{i\alpha\tau A[q_j]}$, $(j = 1, \ldots n; \tau \text{ real})$. Thus, theorem II affirms the uniqueness of the relations that relate the former operators to the latter.

The map $f \to A[f]$, $(f \in \mathcal{F}_L)$, which exists by theorem I, establishes a bijective correspondence between the quantities of Classical Mechanics represented by quadratic polynomials in $p_1, \ldots, p_n$, $q_1, \ldots q_n$, and the quantities of Quantum Mechanics represented by quadratic polynomials formally self-adjoint[t56] in the Schrödinger operators. This correspondence is linear; it correlates classical Poisson brackets and quantum commutators. It is uniquely determined by its precise properties in the form (q_1''), (q_2''). The physical quantities concerned by this correspondence are of fundamental importance; let us cite among them position, momentum, angular momentum, nonrelativistic kinetic energy, potential energy of harmonic oscillators. Let us again note that from an algebraic point of view, this correspondence is an isomorphism between the Lie algebra $\mathcal{F}_L$ of classical quadratic quantities and a Lie algebra of operators in the sense of Segal (section 15). As we proved in the previous section, this correspondence cannot be extended to systems $\mathcal{F}_\Gamma$ of all classical quantities without losing its principal properties.

25 Correspondence for Common Quantities

Even if it is true that the bijective correspondence between classical and quantum quantities that we established for quadratic quantities does not extend to the set of all quantities, it is easy to extend it to all quantities with simple physical significance that we encounter in practice. Indeed, it is well known that for these quantities, replacing the variables $p_1 \ldots, p_n, q_1, \ldots q_n$ with quantum operators does not give rise to any ambiguities that cannot be lifted in a simple and natural way. Let $f(p,q)$ be the element of $\mathcal{F}_\Gamma$ that represents one of these common quantities in Classical Mechanics. In expression (16.9) the operator $\frac{1}{\alpha} H^{(\alpha)}[f]$ associated to f in the representation $\mathcal{R}^{(\alpha)}$ ($\alpha = 2\pi/h$) is

$$\frac{1}{\alpha} H^{(\alpha)}[f] = f(P - P', Q + Q') + \qquad (25.1)$$
$$+ \sum_j \{P'_j f_{p_j}(P - P', Q + Q') - Q'_j f_{q_j}(P - P', Q + Q')\},$$

thus one can easily regroup terms such that the formal cancellation of $P'_1, \ldots P'_n, Q'_1, \ldots Q'_n$ lets the operator (25.1) keep its symmetric character. *Thus, this formal cancellation furnishes a symmetric operator, function of $P_1, \ldots P_n, Q_1, \ldots Q_n$, which is none other than the corresponding quantum operator.* This is what we observe, without any regrouping of terms being necessary, for quantities of the following types: $f = f(p_1, \ldots p_n)$, $f = f(q_1, \ldots q_n)$, $f =$ polynomials of degree 0, 1 or 2 in the p_j, q_j, as well as for their linear combinations. For quantities of types $f = \sum p_j g_j(q_1, \ldots q_n)$, $f = \sum q_j g_j(p_1, \ldots p_n)$, the regrouping of terms is clearly

$$f = \frac{1}{2} \sum \{p_j g_j(q) + g_j(q) p_j\}, \quad f = \frac{1}{2} \sum \{q_j g_j(p) + g_j(p) q_j\}.$$

If $f(p,q)$ is expressed as a Fourier integral, formula (25.1) takes the form (22.6), and the formal cancellation of the P'_j, Q'_j replaces the operator $A(\sigma_1, \ldots \sigma_n, \tau_1, \ldots \tau_n)$ by the identity transformation and takes the operator $\frac{1}{\alpha} H^{(\alpha)}[f]$ into the quantum operator (21.5). Thus, the rule of formal cancellation of $P'_1, \ldots P'_n, Q'_1, \ldots Q'_n$ leads in this case to the Weyl correspondence between classical and quantum quantities.

The correspondence between classical and quantum quantities discussed in the previous section for quadratic quantities in p_j, q_j, has the character of a Lie algebra isomorphism. It loses this character when one extends it to common quantities for which the functional expression is more complicated. Thus, this correspondence makes the subgroup L of Γ play a privileged role. The group T is uniquely related to L: it is the only invariant connected subgroup of L, whereas L is the subgroup of Γ consisting of elements γ such that $\gamma T = T\gamma$. We see that the one-parameter subgroups of T, corresponding to physical quantities position and momentum, are related in the closest possible fashion to one-parameter subgroups of the groups of transformations of spacetime (Galilean or Lorentz group). It is by this special physical significance that T, and therefore L, distinguish themselves in the heart of the group Γ of Classical Mechanics, and this justifies their privileged role in the correspondence between classical and quantum theories.

Translator's Notes

1. This note is just an example in the foreword to the translation.

2. In the text, we replace "infinite group" by "infinite-dimensional group", but we did not want to make that change in the title.

3. A minor typographical point: while the original has ϖ (a different version of π) for this Pfaff form, we are led to believe by other equations that the author intended $\bar{\omega}$ (omega-bar), so we use the latter.

4. The original has "For every element γ of Γ, it suffices to set...".

5. Hypermaximally symmetric and antisymmetric operators are defined in section 2, part c; in brief, a hypermaximally symmetric operator is nowadays called self-adjoint, and an operator A is said to be antisymmetric if iA is symmetric.

6. The original has "continuous infinite".

7. We added "namely the action".

8. We added "The critical points of the action"

9. We added the italics.

10. The original has "considering this group has the flaw of ruining the bijective character of the correspondence mentioned above.". What is elegant in French is not necessarily elegant in English.

11. The original has "Let us briefly indicate how they establish this method in a particular case".

12. The original has "Let us now note the objective of the different chapters."

13. The original has "finite and continuous finite groups".

14. The original has "chapters".

15. We translated *variété linéaire fermée* by "linearly closed manifold". Whereas "manifold" has been part of physics vocabulary for quite a while, the concept of "variety" in English has still not gained widespread acceptance in the physics community, so the author would presumably have used "manifold". By "linearly closed" we understand "closed under linear combination of elements".

16. The original has:
$$\varphi \in D^* \text{ and } \psi = A^*\varphi$$
hence we have for all $\varphi_0 \in \mathcal{D}$
$$(A\varphi_0, \varphi) = (\varphi_0, \psi).$$

17. We have inserted the wedge $\wedge$.

18. The original has "continuously differentiable"

19. We used "rank", although the author was referring to size, irrespective of nullity. The meaning is still clear since the number $2n$ is given.

20. The original has "In the opposite case..." instead of "If there is no such ϵ...".

21. The original has "We will see that this is not always the case."

22. The original has "...actually has no simplicity".

23. We added $m \geq 3$ in this sentence.

24. The original says "we already know that we do not have" and then gives the equation with $\in$ instead of $\notin$.

25. We changed "positive" to "positive-definite".

26. We added " $\iff \tau \in e$ ".

27. The original has "...are different, the same holds for..."

28. We changed the first $\mathcal{H}_{2n}$ to $\mathcal{H}_{2n}^{(\alpha)}$.

29. Again, by "linearly closed" we understand "closed under linear combination of elements".

30. We changed $\mathcal{H}_{2n}$ to $\mathcal{H}_{2n}^{(\alpha)}$.

31. The original has "that (11.2) applies to all its elements, and that $H[f]$ transforms this set onto itself."

32. The original has "defined by"

33. The original has "the real circumference" instead of "a circle".

34. We added the bullets.

35. Our italics.

36. The original has "section 10".

37. We added "respectively"

38. The original had bars on the unprimed m_i in this sentence.

39. The original had bars on the primed m_i in $h_{m_1...m_n}^{m'_1...m'_n}$ and on the unprimed m_i immediately thereafter.

40. The original has p_j as the argument in both operators.

41. The original has no tildes in the argument of φ, nor in the measure.

42. At this point, we would like to recommend the following reference:
    Berry, M., Scaling and nongaussian fluctuations in the catastrophe theory of waves, in Wave propagation and scattering, ed. B.J. Uscinski, Oxford, Clarendon Press (1986) pp. 11-35.
    It contains a discussion of the limit $h \to 0$ in a functional integral perspective, and gives a beautiful (and quite unexpected) application to the reproductive cycle of moth.

43. We translated "easier" (*plus aisée*) to "easier to understand".

44. Our italics.

45. The original has "Uncertainties appearing in the order of factors in products give rise to ambiguities that seem inevitable."

46. The original has "...whether there does not exist..."

47. We added the bullets.

48. Interestingly, while Webster's agrees with *reductio ad absurdum*, the French dictionary Larousse prefers *ab absurdo*. The original uses French, not Latin: *démonstration par l'absurde.*

49. We removed italics in this phrase.

50. The original has "For this point..." rather than "For details...".

51. We added the bullets.

52. The adjoint group is defined on p. 63.

53. The original has Π_j, R, lacking the final j subscript.
54. The original has "example" rather than "counterexample".
55. The original has ϕ instead of the last ψ in equation (24.9).
56. The formal adjoint is obtained by simply exchanging the order of factors Π_j, R_j. See page 75, after eq. (24.7).

References

1. Bargmann, V., Irreducible unitary representations of the Lorentz group, *Ann. Math.* **48** (1947) 568-640.
2. Birkhoff, G., Analytical Groups, *Amer. Math. Soc. Trans.* **43.** (1938) 61-101.
3. Carathéodory, C., Variationsrechnung und partielle Differentialgleichungen erster Ordnung, Leipzig, Teubner (1935).
4. Cartan, E., Leçons sur les invariants intégraux, Paris, Hermann (1922).
5. Chevalley, C., Theory of Lie groups, I, Princeton University Press (1946).
6. Dieudonné, J., Sur les groupes classiques, Paris, Hermann (1948).
7. Dirac, P.A.M., The Principles of Quantum Mechanics, Oxford, Clarendon Press, 2nd Ed (1935).
8. Feynman, R.P., Spacetime approach to nonrelativistic Quantum Mechanics, *Rev. Mod. Phys.* **20** (1948) 367-387.
9. Gelfand, I.M. and Neumark, M.A., Unitary representations of the Lorentz group, *J. Phys. (Moscow)* **10** (1946) 93-94.
10. Gelfand, I.M. and Neumark, M.A., Unitary representations of the Lorentz group. In Russian, *Bull. Acad. Sci. USSR., Math. Ser.* **11** (1947) 411-504.
11. Godement, R., Théorie générale des sommes continues d'espaces de Banach, *C.R. Acad. Sci. Paris* **t.228** (1949) 1321-1323.
12. Goursat, E., Leçons sur le problème de Pfaff, Paris, Hermann (1922).
13. Groenewold, H.J., On the principles of elementary quantum mechanics, The Hague, Nijhoff (1946).
14. Hopf, E., Ergodentheorie, *Erg. d. Math.* **5, no. 2.** (1937)
15. Koopman, B.O., Hamiltonian systems and linear transformations in Hilbert space, *Proc. Nat. Acad. Sci. USA* **17** (1931) 315.
16. La Vallée Poussin, C. de, Intégrales de Lebesgue, fonctions d'ensemble, classes de Baire, Paris, Gauthier-Villars (1934).
17. Lefschetz, S., Introduction to topology, Princeton University Press (1949).
18. Mautner, F.I., Unitary representations of locally compact groups I, *Ann. Math.* **51** (1950) 1-25.
19. Neumann, J. von, Allgemeine Eigenwerttheorie Hermitischer Funktionaloperatoren, *Math. Ann.* **102** (1929) 49-131.
20. Neumann, J. von, Zur Algebra der Funktionaloperatoren und Theorie der normalen Operatoren, *Math. Ann.* **102** (1929) 370-427.
21. Neumann, J. von, Über Funktionen von Funktionaloperatoren, *Ann.*

Math. **32** (1931) 191-226.
22. Neumann, J. von, Die Eindeutigkeit der Schrödingerschen Operatoren, *Math. Ann.* **104** (1931) 570-578.
23. Neumann, J. von, Über einen Satz von Herrn M. H. Stone, *Ann. of Math.* **33** (1932) 567-573.
24. Neumann, J. von, Zur Operatormethode in der klassischen Mechanik, *Ann. Math.* **33** (1932) 587-642.
25. Neumann, J. von, On rings of operators, reduction theory, *Ann. Math.* **50** (1949) 401-485.
26. Pfaff, G.F., Methodus generalis, æquationes differentiarum partialum, necnon æquationes differentiales vulgares, utrasque primi ordinis, inter quotcumque variabiles, complete integrandi, *Abh. K. Preuss. Akad. Wiss. Berlin* (1814) 76-136.
27. Riesz, F., Über die lineare Transformationen des komplexen Hilbertschen Raumes, *Acta Litt. Ac. Sc. Szeged* **5** (1930) 19-54.
28. Segal, I.E., A class of operator algebras which are determined by groups, *Duke Math. Journ.* **18** (1951) 221-265.
29. Stone, M.H., Linear transformations in Hilbert space. Operational methods and group theory, *Proc. Natl. Acad. Sci. USA* **16** (1930) 172-175.
30. Stone, M.H., Linear Transformations in Hilbert Space and their Applications to Analysis, *Amer. Math. Soc. Coll. Publ., vol. XV, New York.* (1932)
31. Stone, M.H., On one-parameter unitary groups in Hilbert space, *Ann. Math.* **33** (1932) 643-648.
32. VanHove, L., A set of unitary representations of the group of contact transformations, *Proc. Intl. Congress of Mathematicians, Cambridge, USA, in press (1950)*.
33. Weil, A., L'intégration dans les groupes topologiques et ses applications, Paris, Hermann (1940).
34. Weyl, H., Quantenmechanik und Gruppentheorie, *Z. Phys.* **46** (1928) 387-395.
35. Whittaker, E.T., A treatise on the analytical dynamics of particles and rigid bodies, Cambridge, University Press (1927).

Abstract

The group of canonical transformations of the $2n$-dimensional (p,q) space is enlarged to the group Γ of smooth bijective transformations of the $(2n+1)$-dimensional space (s,p,q) that leave invariant the one-form $ds - \sum_1^n p_j dq_j$. The group of canonical transformations is the quotient group Γ/C of Γ by its center C. The group Γ also serves for constructing the symplectic group and the group of homogenous canonical transformations. The set of infinitesimal transformations in Γ, called $X[\mathcal{F}_\Gamma]$, cannot always be given a Lie algebra structure since Γ is an infinite-dimensional group and the space (s,p,q) is noncompact. This set is in bijective correspondence with a family $\mathcal{F}_\Gamma$ of functions f satisfying the differential equations (5.1), (5.2). With further requirements on f, the author constructs a family $\mathcal{F} \subset \mathcal{F}_\Gamma$ and a Lie algebra isomorphism between $\mathcal{F}$ and the set $X[\mathcal{F}]$. A unitary representation $\mathcal{R}$ of Γ on the Hilbert space $\mathcal{H}_{2n+1}$ of functions ϕ on (s,p,q) is the continuous sum of representations $\mathcal{R}^{(\alpha)}$ of Γ on the Hilbert spaces $\mathcal{H}_{2n}^{(\alpha)}$ of functions $\varphi^{(\alpha)}$ on (p,q). A representation $\mathcal{R}^{(\alpha=0)}$ is a representation of Γ/C. The author constructs the anti-selfadjoint operators in $\mathcal{R}$, $\mathcal{R}^{(\alpha)}$, identifies their domains, and proves that they form an operator Lie algebra in the sense of Segal. A unitary transformation W of $\mathcal{H}_{2n}$ transforms these operators into operators acting only on n of the $2n$ independent variables $\{p_j, q_j\}$. The goal of this transformation is a comparison between quantum mechanics and classical mechanics in a Hilbert space setting.

ECOLE d'ÉTÉ de PHYSIQUE THÉORIQUE
1235

UNIVERSITE DE GRENOBLE.

ECOLE D'ETE DE PHYSIQUE THEORIQUE.

COURS DE MECANIQUE QUANTIQUE.

Léon Van Hove

Associé du Fonds National de la Recherche Scientifique de Belgique,
Assistant à l'Université libre de Bruxelles.

TABLE DES MATIERES.

Avertissement 1

CHAPITRE I : Généralités sur la description quantique d'un système.
I.1 - Représentation d'une particule ponctuelle dans la théorie de Schrödinger 2
I.2 - Description quantique des états et des grandeurs physiques pour un système quelconque 3
I.3 - Grandeurs simultanément mesurables 6
I.4 - Grandeurs non simultanément mesurables - Rôle du commutateur 7
I.5 - Combinaison de degrés de liberté - Produit direct d'espaces de Hilbert 8
I.6 - Groupe engendré par un opérateur hypermaximal symétrique 9
I.7 - Relations de commutation de Heisenberg 9
I.8 - Evolution temporelle d'un système 10
I.9 - Mouvement d'un paquet d'onde - Cas limite de la Mécanique classique 11

CHAPITRE II : L'équation de Schrödinger et la diffusion d'une particule par une force centrale.
II.1 - Nature du spectre d'énergie 15
II.2 - Potentiel à symétrie sphérique 16
II.3 - Etats stationnaires du spectre continu en l'absence de potentiel coulombien 16
II.4 - Décomposition d'une onde plane en ondes sphériques 20
II.5 - Section efficace de diffusion 21
II.6 - L'équation de Schrödinger dans l'espace des impulsions 23

CHAPITRE III : Moments angulaires.
III.1 - Groupe des rotations de l'espace à trois dimensions et groupe unitaire unimodulaire à deux variables 24
III.2 - Représentations linéaires du groupe des rotations de l'espace euclidien réel à trois dimensions 25
III.3 - Moments angulaires en Mécanique quantique 27
III.4 - Composition des moments angulaires 27

CHAPITRE IV : L'équation de Klein-Gordon.
IV.1 - L'équation de Klein-Gordon 30
IV.2 - Solutions singulières 31
IV.3 - Fonctions de Green 32
IV.4 - Représentation de Fourier des fonctions de Green 35

CHAPITRE V : L'équation de Dirac.
V.1 - Construction de l'équation de Dirac 37
V.2 - Invariance relativiste de l'équation de Dirac 39
V.3 - L'expression usuelle des matrices de Dirac - Le spin 41
V.4 - Conjugaison de charge 42
V.5 - Ondes planes - Approximation non relativiste 43
V.6 - Paradoxe de Klein 44

CHAPITRE VI : Méthodes de perturbation.
VI.1 - Perturbations des niveaux d'énergie et des états stationnaires . . 45
VI.2 - Transitions causées par une perturbation 46
VI.3 - Approximation de Born 49

CHAPITRE VII : L'opération de mesure en Mécanique quantique.
VII.1 - Mesure d'une grandeur et réduction du paquet d'onde 51
VII.2 - Discussion du comportement de l'appareil de mesure 53
VII.3 - Détermination de la fonction d'onde par des mesures 55

Avertissement.

La grande majorité des sujets abordés dans ces leçons sont classiques et divers traités en donnent d'excellents exposés. Nous aurons donc souvent à nous occuper de questions familières à la plupart des auditeurs. Nous rappellerons très brièvement leurs aspects fondamentaux de façon à pouvoir nous étendre sur des aspects moins bien connus, souvent liés aux développements récents de la Physique quantique. Parfois aussi nous developperons un peu plus qu'on ne le fait d'habitude les diverses méthodes mathématiques auxquelles la théorie quantique fait appel ; dans aucun cas cependant nous ne nous préoccuperons de rigueur mathématique.

Les présentes notes sont assez condensées. Les raisonnements et les calculs ne seront developpés en détail que dans le cours oral.

Parmi les ouvrages de base qui seront utilement consultés, citons :

W. Pauli, Die allgemeinen Prinzipien der Wellenmechanik, Handb. der Physik, Bd XXIV$_1$;

P.A.M. Dirac, The Principels of Quantum Mechanics, Oxford, Clarendon Press;

N.F. Mott and H.S.W. Massey, The Theory of atomic Collisions, Oxford, Clarendon Press ;

E.C. Kemble, The Fundamental Principles of Quantum Mechanics, New York, Mc Graw Hill ;

L.I. Schiff, Quantum Mechanics, New York, Mc Graw Hill;

N.F. Mott and I.N. Sneddon, Wave Mechanics and its Applications, Oxford, Clarendon Press.